Evaluation of double recirculating aquaponic systems for intensive plant production

Evaluation of double recirculating aquaponic systems for intensive plant production

Dissertation

zur Erlangung des akademischen Grades
doctor rerum agriculturarum
(Dr. rer. agr.)

eingereicht an der
Lebenswissenschaftlichen Fakultät
der Humboldt-Universität zu Berlin

von

Master of Science Johanna Suhl

Geboren am 15.05.1984 in Gifhorn

Präsidentin der Humboldt-Universität zu Berlin: Prof. Dr.-Ing. Dr. Sabine Kunst
Dekan der Lebenswissenschaftlichen Fakultät: Prof. Dr. Bernhard Grimm

Gutachter:

1. Prof. Dr. Uwe Schmidt
2. Dr. Klaus Knopf
3. Prof. Em. Gerrit van Straten

Tag der mündlichen Prüfung: 09.12.2019

Bibliografische Information der Deutschen Nationalbibliothek
Die Deutsche Nationalbibliothek verzeichnet diese Publikation in der Deutschen Nationalbibliografie; detaillierte bibliographische Daten sind im Internet über http://dnb.d-nb.de abrufbar.
1. Aufl. - Göttingen: Cuvillier, 2020
Zugl.: Berlin, Humboldt-Univ., Diss., 2019

Nonnenstieg 8, 37075 Göttingen
Telefon: 0551-54724-0
Telefax: 0551-54724-21
www.cuvillier.de

1. Auflage, 2020
Gedruckt auf umweltfreundlichem, säurefreiem Papier aus nachhaltiger Forstwirtschaft.

ISBN 978-3-7369-7159-2
eISBN 978-3-7369-6159-3

List of publications

2014

Dannehl, D., **Suhl, J.**, Huyskens-Keil, S., Ulrichs, C. & Schmidt, U. (2014). Effects of a special solar collector greenhouse on water balance, fruit quantity and fruit quality of tomatoes. *Agricultural Water Management, 134, 14–23.*

Schulz, J., Berk, J., **Suhl, J.**, Schrader, L., Kaufhold, S., Mewis, I., Hafez, H. M. & Ulrichs, C. (2014). Characterization, mode of action, and efficacy of twelve silica-based acaricides against poultry red mite (*Dermanyssus gallinae*) in vitro. *Parasitology Research, 1–9*

2015

Dannehl, D., **Suhl, J.,** Ulrichs, C., & Schmidt, U. (2015). Evaluation of substitutes for rock wool as growing substrate for hydroponic tomato production. *Journal of Applied Botany and Food* Quality, 88(1).

Kloas, W., Groß, R., Baganz, D., Graupner, J., Monsees, H., Schmidt, U., Staaks, G., **Suhl, J.**, Tschirner, M., Wittstock, B., Wuertz, S., Zikova, A. & Rennert, B. (2015). A new concept for aquaponic systems to improve sustainability, increase productivity, and reduce environmental impacts. *Aquaculture Environment Interactions, 7(2), 179–192.*

2016

***Suhl, J.**, Dannehl, D., Kloas, W., Baganz, D., Jobs, S., Scheibe, G., & Schmidt, U. (2016). Advanced aquaponics: Evaluation of intensive tomato production in aquaponics vs. conventional hydroponics. *Agricultural Water Management, 178, 335–344.*

Dannehl, D., Becker, C., **Suhl, J.**, Josuttis, M., & Schmidt, U. (2016). Reuse of organo-mineral substrate waste from hydroponic systems as fertilizer in open-field production increases yields, flavonoid glycosides, and caffeic acid derivatives of red oak leaf lettuce (*Lactuca sativa* L.) much more than synthetic fertilizer. *Journal of Agricultural and Food Chemistry, 64(38), 7068–7075.*

2018

***Suhl, J.,** Dannehl, D., Baganz, D., Schmidt, U. & Kloas, W. (2018). An innovative suction filter device reduces nitrogen loss in double recirculating aquaponic systems. *Aquaculture Engineering, 82, 63–72.*

---

*These publications are included in the present thesis

***Suhl, J.,** Dannehl, D., Zechmeister, L., Baganz, D., Kloas, W., Lehmann, B., Scheibe, G. & Schmidt, U. (2018) Prospects and challenges of double recirculating aquaponic systems (DRAPS) for intensive plant production. *Acta Horticulturae, 1227, 449–456.*

2019

***Suhl, J.,** Oppedijk, B., Baganz, D., Kloas, W., Schmidt, U. & van Duijn, B. (2019). Oxygen consumption in recirculating nutrient film technique in aquaponics. *Scientia Horticulturae, 255, 281–291.*

***Suhl, J.,** Baganz, D., Kloas, W., Dannehl, D., Jobs, S., Scheibe, G. & Schmidt, U. (2019). The potential of double recirculating aquaponic systems for intensive tomato production. *Acta Horticulturae, 1242, 679–686.*

*Monsees, H.♦, **Suhl, J.**♦, Paul, M., Kloas, W., Dannehl, D., & Würtz, S. (2019). Lettuce (*Lactuca sativa*, variety Salanova) production in decoupled aquaponics systems reveals same yield and similar quality as in conventional hydroponic systems but can reduce drastically greenhouse gas emission indirectly by saving production of fertilizer. *PLoS ONE, 14(6): e0218368.* (♦These authors contributed equally to this work)

Conference contribution (in terms of aquaponics)

Oral presentations

2015

Suhl, J. & Schmidt, U. (2015). Nährstoffdynamik und Pflanzenmonitoring in Aquaponik-Systemen. 8. Agrosnet Doktorandentag in Berlin, Germany

Suhl, J., Baganz, D., Kloas, W., Dannehl, D. & Schmidt, U. (2015). A double recirculating aquaponic system and its impact on sustainable food production. The 4th International Research Conference on Basic and Agriculture Science – Universidad Autónoma de Chapingo, Mexico

2016

Suhl, J., Kloas, W. & Schmidt, U. (2016). Lebensmittelproduktion auf begrenzten Flächen: Neue Aquaponik-Technologie für das „combined farming“ – Gemüsebau und Fischzucht in Intensivproduktionssystemen. Kick-Off-Workshop FoodBerlin, Germany

Suhl, J., Baganz, D., Jobs., S., Scheibe, G., Kloas, W. & Schmidt, U. (2016). Innovative aquaponics for professional application (INAPRO). 3rd International Symposium on Horticulture in Europe (SHE) in Chania, Greence

2017

Suhl, J., Zechmeister, L., Dannehl, D., Baganz, D., Kloas, W., Lehmann, B., Jobs, S., Scheibe, G. & Schmidt, U. (2017). Prospects and challenges of double recirculating aquaponic systems (DRAPS) for intensive tomato production. International Symposium on New Technologies for Environment Control, Energy-Saving and Crop Production in Greenhouse and Plant Factory (Greensys) in Beijing, China

Suhl, J., Dannehl, D., Baganz, D., Kloas, W. & Schmidt, U. (2017). Aquaponics and its potential for intensive crop production. International Workshop: Urban Horticulture in Controlled Environments – Universidad Autónoma de Chapingo, México

Poster

Suhl, J., Baganz, D., Staaks, G., Andrä, J., Dannehl, D., Schmidt, U. & Kloas, W. (2016). Nährionendynamik in einem innovativen, doppelt rezirkulierenden Aquaponik-System. Deutscher Fischereitag in Rostock, Germany

Suhl, J., Dannehl, D., Baganz, D., Kloas, W., Lehmann, B., Scheibe, G. & Schmidt, U. (2017). Perspektiven und Herausforderungen doppelt rezirkulierender Aquaponiksysteme (DRAPS) für die intensive Pflanzenproduktion. 51. Gartenbauwissenschaftliche Tagung der DGG und des BHGL

Suhl, J., Dannehl, D., Baganz, D., Kloas, W., Lehmann, B., Scheibe, G. & Schmidt, U. (2018) Nutrient concentrations in fish waste water and its potential to use it for tomato production in double recirculating aquaponic systems. 52. Gartenbauwissenschaftliche Tagung der DGG und des BHGL

Abctract

The development of double recirculating aquaponic systems (DRAPS) presents new possibilities and initiates a new era of aquaponics. DRAPS combines two unidirectional connected and independent recirculating units – a recirculating aquaculture system (RAS) for fish production and a recirculating hydroponic cycle for plant production. This enhancement offers a solution for the limitations of single recirculating aquaponic systems (SRAPS). In SRAPS three different biological systems (fish, plants, and nitrifying bacteria) with different requirements must be merged in a one loop system. Due to the different environmental requirements of fishes and of plants, compromises in terms of water quality have to be made. This in turn can result in reduced yields compared to separate single production systems for fishes and plants, respectively. In contrast, due to its independent operating cycles, DRAPS allow the production of fish and plants under specific optimal conditions.

The present thesis was done in the framework of the EU INAPRO project (Innovative Aquaponics for Professional Application). The thesis is based on three main experiments, production of Nile tilapia or African catfish in combination with tomatoes or lettuce. The research focused primarily on the hydroponic unit. One major subject was to evaluate the DRAPS as a whole system during a whole year tomato production cycle. Tomato and lettuce production in DRAPS was compared to that of conventional hydroponic conditions (control). As such, quantity and quality aspects of the vegetables were investigated. The direct comparison was especially important. DRPAS was conceived as an intensive food production system and was developed as alternative to conventional hydroponics. This means it has to compete to conventional intensive vegetable production however very few studies are available comparing DRAPS and conventional hydroponics. Another focus of the present thesis was to study the connection point between fish and plant in more detail and to identify possible weaknesses of this point. Furthermore, the effect of using fish waste water on the oxygen concentration in the nutrient solution was investigated. To evaluate the sustainability of DRAPS, the fertiliser and fresh water saving potential were investigated. Additionally, how much emissions (CO_2-equivaltence) could be saved by replacing nitrogen fertiliser by fish waste water use was calculated.

The results of the present thesis demonstrate clearly the competiveness of DRAPS with conventional hydroponic systems in terms of plant production. In both years the tomato yields were not significant. The mean yield of tomatoes produced in DRAPS was 26.5 kg m^{-2} (mean value of two production years) while conventional hydroponics was 29.0 kg m^{-2}. These results were independent of whether Nile Tilapia or African catfish was produced. The fresh weight of lettuce, irrigated with Nile Tilapia waste water, did also not differ significantly between DRAPS and conventional hydroponics. The mean fresh weight of lettuce produced in DRAPS was 323.3 g in comparison to 325.9 g in conventional hydroponics. For the lettuce experiment one of the fish waste water based nutrient solution was disinfected by heating. It was found that the microbiology of the fish waste water did not have a significant effect on lettuce growth. In terms of quality parameters, like sugar/acid ratio or secondary metabolites, neither the tomato fruit quality nor the lettuce quality was affected by the conditions of DRAPS. However, it was found that the oxygen concentration in the cultivation trenches can drop to critical levels when nutrient solution is prepared with fish waste water. But no negative effects on plant performance were detected in the corresponding study. The oxygen depletion was probably promoted by the original construction of the research DRAPS using a 3-champer pit (3-cp).

Using DRAPS was shown to reduce the fresh water and fertiliser use. Due to the use of fish waste water, the fresh water demand only for the hydroponic unit was in principle zero. Due to the similar biomass output (between DRAPS and control) in the hydroponic unit, it can be concluded that in the complete DRAPS the total biomass output per unit fresh water was increased due to the fish biomass gained in the recirculating aquaponic system. Due to results within the duration of the project a modification of the original research DRAPS was conducted. Due to the modification the nitrogen loss, originated in the 3-cp of the original DRAPS, were reduced significantly and the fertiliser saving was increased. The saving potential for mineral fertiliser in DRAPS was above 60%. This in turn can contribute immensely to environmental protection. Mineral based fertiliser manufacturing, especially the production of nitrogen fertiliser is a highly energy demanding process. Consequently, just due to the reduced mineral fertiliser requirement for DRAPS, the environment can be preserved.

Despite the very favourable results of the present thesis, the research on DRAPS is still ongoing. Especially the connection point between the fish and plant unit is still a bottleneck of DRAPS. Recommendation of DRAPS as an easy to handle system for a practical approach as intensive large-scale food production cannot be given. Nevertheless, DRAPS is a highly promising system which must be investigated and improved further. Finally, it will contribute significantly to sustainable intensification and consequently to food safety.

Zusammenfassung

Die Entwicklung von doppelt rezirkulierenden Aquaponiksystemen (DRAPS) bietet neue Möglichkeiten und eröffnet eine neue Ära in der Aquaponik. DRAPS kombinieren zwei unabhängig voneinander rezirkulierende Systeme: ein rezirkulierendes Aquakultursystem (RAS) für die Fischproduktion und ein rezirkulierendes Hydroponiksystem für die Produktion von Pflanzen. Die Unabhängigkeit beider Systeme bietet eine Lösung für das Problem, welches bei einfach rezirkulierenden Aquaponiksystemen (SRAPS) auftritt. Die Herausforderung bei SRAPS ist, dass drei unterschiedliche biologische Systeme (Fische, Pflanzen und nitrifizierende Bakterien) mit unterschiedlichen Ansprüchen in einem System zusammengebracht werden müssen. In Hinblick auf die Wasserqualität werden daher Kompromisse eingegangen und die unterschiedlichen biologischen Systeme befinden sich nicht im Optimum. Das wiederum kann dazu führen, dass im Vergleich zu separaten, konventionellen Fisch- oder Pflanzenproduktionssystemen, die Erträge oftmals geringer ausfallen. Im Gegensatz dazu erlauben DRAPS die Einstellung optimaler Konditionen für die zu produzierenden Fische beziehungsweise Pflanzen.

Die aktuelle Doktorarbeit wurde im Rahmen des EU INAPRO Projektes (Innovative Aquaponics for Professional Application) durchgeführt. Die Arbeit basiert auf drei Hauptexperimenten in denen entweder Nil Tilapia oder Afrikanischer Raubwels jeweils in Kombination mit Tomaten oder Salat produziert worden ist. Ein Schwerpunkt der Arbeit war es, DRAPS als ganzheitliches System während eines gesamten Jahreszyklus von Tomate zu evaluieren. Dabei lag der Fokus auf der Hydroponik. Die Produktion von Tomaten und Salat in DRAPS wurde dabei mit der konventionellen Produktion, also in konventionellen rezirkulierenden Hydroponikanlagen (Kontrolle), verglichen. Vor allem der Einfluss von DRAPS auf die quantitativen und qualitativen Pflanzenparameter ist untersucht worden. Der direkte Vergleich von DRAPS mit der konventionellen Hydroponik war eine absolute Notwendigkeit. Auf der einen Seite wurden DRAPS für die intensive Lebensmittelproduktion entwickelt und müssen daher mit konventionellen Produktionssystemen konkurrieren können. Zum anderen sind bisher nur sehr wenige empirische Studien bekannt, die sich mit diesem direkten Vergleich beschäftigen. Ein weiterer Aspekt der vorliegenden Arbeit war es, die Verbindungseinheit zwischen dem RAS und der Hydroponik tiefergehend zu untersuchen und mögliche Schwächen zu identifizieren. Weiterhin ist untersucht worden, welchen Effekt die Fischabwassernutzung auf den Sauerstoffgehalt in der damit angemischten Nährlösung hat. Um die Nachhaltigkeit des Systems im Vergleich zu konventionellen Systemen darzustellen wurde das Frischwasser- und Düngeeinsparungspotential von DRAPS kalkuliert. Weiterhin ist berechnet worden, wie viele Emissionen (CO_2-equivaltene) eingespart werden können, wenn durch die Nutzung von Fischabwasser die Menge an mineralischen Stickstoffdünger reduziert werden kann.

Die Ergebnisse dieser Arbeit zeigen deutlich, dass DRAPS in Bezug auf Pflanzenquantität und –qualität konkurrenzfähig zur konventionellen Hydroponikproduktion sind. Die Tomatenerträge unterschieden sich, unabhängig von der produzierten Fischart (Nil Tilapia oder Afrikanischer Raubwels), in den einzelnen Versuchsjahren nicht signifikant voneinander. Während der Durchschnittsertrag in DRPAS aus zwei Produktionsjahren bei 26,5 kg m^2 lag, wurden in der konventionellen Hydroponik im Durchschnitt von zwei Jahren 29,0 kg m^2 Tomaten geerntet. Die Frischmasse der Salatköpfe, die mit Fischabwasser vom Nil Tilapia versorgt wurden, unterschied sich ebenso nicht signifikant. Während die Salatköpfe in der DRAPS-Variante durchschnittlich 323,3 g wogen, brachten die Salatköpfe aus der Kontrollvariante 325,9 g auf die Waage. Für das Salat-experiment wurde für eine Variante das Fischabwasser thermisch desinfiziert. Es zeigte sich, dass die Mikrobiologie im Fischabwasser keinen signifikanten Einfluss auf das Wachstum der Salatpflanzen hatte. In Bezug auf die Qualität der Gemüsearten, zum Beispiel das Zucker-Säure-Verhältnis oder verschiedenen Sekundärmetaboliten, konnte ebenfalls kein eindeutiger Effekt durch die Produktion der Pflanzen in DRAPS festgestellt werden. Allerdings ist festgestellt worden, dass der Sauerstoffgehalt in der Nährlösung die mit Fischwasser angemischt wurde, auf kritische Konzentrationen absinken kann. Allerdings konnten in der vorliegenden Studie keine negativen Effekte auf das Pflanzenwachstum festgestellt werden. Die Sauerstoffabnahme resultierte vermutlich aus der Bauweise des originalen DRAPS und der eingebauten 3-Kammergrube. Der Frischwasser- und Düngerbedarf von DRAPS waren deutlich gegenüber der konventionellen Hydroponik reduziert. Durch die Verwendung von Fischabwasser war der Frischwasserverbrauch, separat für die Hydroponik betrachtet, gleich null. Durch die gleichen Gemüseerträge erhöhte sich, betrachtet für das Gesamtsystem, im Vergleich zur konventionellen Hydroponik, die Biomasseproduktion je Wassereinheit folglich um die Fischerträge. Aufgrund erster Ergebnisse während des Projektes wurde das originale DRAPS modifiziert und umgebaut. Dadurch konnte der Stickstoffverlust, der in der 3-Kammergrube festgestellt wurde, signifikant reduziert werden und die Düngereinsparung verbessert werden. Das Einsparungspotential für Mineraldünger betrug über 60 %. Damit können DRAPS zur Entlastung der Umwelt beitragen. Die Produktion von Mineraldüngern, insbesondere Stickstoffdünger, weist einen sehr hohen Energiebedarf auf und verursacht Emissionen. Allein durch die reduzierte Produktion von Mineraldüngern beziehungsweise Stickstoffdünger können DRAPS zu einer Schonung der Umwelt beitragen.

Trotz der guten und vielversprechenden Ergebnisse macht diese Arbeit auch deutlich, dass weitere Forschungsaktivitäten in Bezug auf DRAPS notwendig sind. Insbesondere die Verbindungseinheit zwischen Fisch- und Pflanzenproduktion stellt weiterhin eine Herausforderung dar. Eine bedenkenlose Empfehlung für die praktische Anwendung von DRAPS als umfangreiches, intensives Lebensmittelproduktionssystem kann noch nicht ausgesprochen werden. Nichtsdestoweniger, DRAPS ist ein absolut vielversprechendes System, welches weiter vorangebracht werden muss. Dann kann es einen Beitrag zur nachhaltigen Intensivierung und damit zur Lebensmittelsicherheit beitragen.

Contents

List of tables

List of figures

List of formulars

List of abbreviations

%	percentage
°C	degree centigrade
3-cp	3-chamber pit
B	boron
BER	blossom end rot
Ca	calcium
CFA	continuous flow analysis
Chl NDI	Remittance (R) spectra indices for chlorophyll content
cm	centimetre
CO_2	carbon dioxide
$CO_{2\text{-e}}$/$CO_{2\text{-eq}}$	carbon dioxide equivalent
Cu	copper
CW	calendar week
DI	distilled
DM	dry matter
DRAPS	double recirculating aquaponic system
dS	deci siemens
EC	electrical conductivity
Eq	Equation
FAO	Food and Agriculture Organization of the United Nations
FCR	feed conversion ratio (kg feed input divided by total weight gain)
Fe	iron
FUE	total fertiliser use efficiency
FW	fresh weight
FWUE	total fresh water use efficiency
HCl	hydrochloric acid
ICP-OES	inductively coupled plasma-optical emission spectrometry
INAPRO	Innovative aquaponics for professional application (EU project)
K	potassium
kg	kilogram
L	litre
LA	leaf area
LN	number of leaves
m	metre

m^2	square metre
m^3	cubic metre
Mg	magnesium
mg	milligram
min	minute
mL	millilitre
Mn	manganese
Mo	molybdenum
N	nitrogen
Na	sodium
NCDs	chronic non-communicable diseases
NH_3	ammonia
NH_4^+	ammonium
NH_4-N	ammonium nitrogen
NH_4NO_3	ammonium nitrate
N_{min}	sum of NO_3^--N and NH_4^--N
NO_2^-	nitrite
NO_3^-	nitrate
NO_3^--N	nitrate nitrogen
NT	nutrient solution tank
N_{total}	sum of NO_3^--N and NH_4^--N
OECD	Organization for Economic Co-operation and Development
P	phosphorus
RAS	recirculating aquaculture systems
S	sulphur
Sx-x	supporting material (x indicates numbers)
SAR	sugar-acid ratio
SRAPS	single recirculating aquaponic system
SSC	soluble solids content
SWOT	strengths, weaknesses, opportunities, and threats
TA	titratable acid
UV	ultraviolet
W	watt
WHO	World Health Organization
Zn	zinc

Nomenclature and descriptions

To make the content of the present thesis clearer, some specific used words and synonymous used words are described here.

Synonymous used words:

- Double recirculating aquaponic systems (DRAPS) are the same as decoupled aquaponics
- Single recirculating aquaponic systems (SRAPS) are the same as one-loop aquaponics
- Recirculating hydroponic systems is also referred to as closed hydroponic systems

Meaning/Description of some words:

Word	Meaning/Description
Fish waste water	Is the water that is used for plant irrigation, after it is removed from the recirculating aquaculture system (RAS)
Fish water	Is the water that is recirculating in the RAS
INAPRO* research DRAPS	Small research DRAPS (196 m^2) build as part of the EU INAPRO project. Used for experiment 1 and 2 of the present study.
INAPRO demonstration DRAPS	Bigger demonstration DRAPS ($\approx$ 550 m^2) build as part of the EU INAPRO project.
Nutrient solution	Is the finally prepared nutrient solution to irrigate the plants
Stock solutions	Are the concentrated nutrient stock solutions with high salt content, used to prepare the nutrient solution (Subsection 6.3.2.2, Figure 14)
Light version of DRAPS	Without active cooling unit in the greenhouse
Full version of DRAPS	With active cooling unit in the greenhouse to recondensate the evapotranspirated water of the plants

*INAPRO = Innovative aquaponics for professional application (EU project)

1

General introduction

1 General introduction

1.1 Scientific background

Agriculture is sitting on the fence. On the one hand there is a growing world population which needs to be fed, and on the other hand agriculture is affected by climate change but also water and energy scarcity. The world´s population is predicted to rise to nearly nine to ten billion by 2050 (worldometers.info). Therefore the overall food production has to increase by 70% calculated from 2005/07 to 2050 (FAO, 2009). At the same time agriculture is responsible for 24% of the global greenhouse gas emissions, mainly CH_4 and N_2O (Edenhofer et al., 2014) and withdrawal of nearly 70% of the global fresh water resources (FAO, 2016a). Intensive food production systems like vegetable production in greenhouses and aquaculture are highly productive on the one hand, but also have a significant negative impact on the environment. However, the current significant research efforts and developments have been made to make these systems more environmentally friendly. Sustainable food production systems with low resource requirements, and simultaneously high biomass output, become more and more important for the future and is referred to as sustainable intensification (Godfray et al., 2010).

One suitable option for this is the relatively new technology of double recirculating aquaponic system (DRAPS) – a high-tech system for intensive, but more sustainable food production. Generally, aquaponics uses the synergetic effects of the combined production of fish and plants. Conventional single recirculating aquaponic systems (SRAPS) are also referred to as one-loop systems are low-tech aquaponics. It is an old approach to use nutrients and water twice by producing two cash crops. As such, the nutrient-rich fish water is used to irrigate and feed the hydroponically grown plants, while the fish water itself is simultaneously cleaned from fish toxic ammonium. The functionality of SRAPS is based on the interaction of three different biological systems, fish, bacteria and plants. But the associated challenge, the different environmental requirements of the three species, prevented aquaponics from its implementation as a key player for large-scale production units, so far.

Recently, aquaponics was rethought and a new concept was developed to solve this challenge. This new, high-tech aquaponic system is based on two separate and independent working recirculating systems. As such, a recirculating aquaculture system (RAS) and a recirculating (closed) hydroponic system are connected unidirectionally. Independent working means that both units can generally operate completely without each other. Certainly, the fish water is still used to irrigate and feed the plants, but the water from the plants is not delivered back to the fish directly. This provides the opportunity for optimisation of fish water and adjustment specifically to the requirements of the respective plants. Mineral fertiliser and acid is added to adjust the nutrient concentration or the pH value. This, in turn, means that the fish, as well as the plants, can be produced under optimal and intensive conditions. These systems are referred to as double recirculating aquaponic system (DRAPS) or decoupled aquaponics. The concept of DRAPS is relatively new and the current state of research is given in section 2.3.2.

1.2 Delimitation, aims, and objectives of the present study

1.2.1 Delimitation to already published theses

Recently two doctoral theses dealing more or less with DRAPS were published. The titles are "Opportunities and Challenges of Multi-Loop Aquaponic Systems" written by Simon Goddek (2017) and "Overcoming major bottlenecks in aquaponics – A practical approach" written by Hendrik Monsees (2018). A third thesis studied DRAPS only at the edge and was written by Delaide (2017). However, his main subject was on single recirculating aquaponics, and in terms of DRAPS, he worked closely together with Goddek (2017). Therefore Delaides thesis will not be considered at this place. The already published theses might raise the question if a further thesis can contribute increasing the knowledge in terms of DRAPS. In the following is demonstrated why the present thesis has its justification.

The major topics of the different theses differed or even came to different results which are discussed in the present thesis. Goddek (2017) worked in his thesis with smaller experimental setups and with technical approaches, where he modelled aquaponic systems with different technical components. Briefly, his experimental setups were conducted in separated hydroponics systems using the "idea" of DRAPS. He used fish water to irrigate the plants but he did not work in a whole operating DRAPS. In another experimental setup, he investigated the remobilization of nutrients from sludge under anaerobic conditions. Additionally, Goddek (2017) included two technical chapters where he designed and modelled a multi-loop aquaponic system including a further anaerobic nutrient remobilization component or desalination technology.

In contrast, Monsees (2018) worked partly in a whole operating double recirculating aquaponic system (in the prototype of DRAPS developed and build by Rennert et al. (2011) and Kloas et al. (2015), slightly modified). His thesis is divided into three main sections. Initially, he identified which nitrate concentration Nile tilapia can be produced in RAS without negatively affecting health and growth. In his 2nd experiment, he compared DRAPS and SRAPS under real conditions with the DRAPS prototype in terms of fish production and plant growth. The last main chapter deals also with sludge treatment for nutrient remobilisation like Goddek (2017). The results of both are further discussed in subsection 2.3.2.

Besides the valuable studies of both, there is a missing proof of concept on an intensive production basis. Even if Monsees (2018) worked already with a whole DRAPS, comparing DRAPS with low-tech single recirculating aquaponic systems (SRAPS). But for the implementation as large-scale or intensive food production system, the comparison to conventional hydroponically single production is absolutely necessary. This was compared already by Goddek (2017). However, he conducted relatively small scale experiments without using a whole operating DRAPS. Additionally, he did not evaluate a whole DRAPS during a whole year production cycle. Therefore, one major subject of the present thesis was to evaluate an operating DRAPS as a whole system during a whole year production cycle in comparison to conventional hydroponics. Besides this, some further subjects, which were not considered by the mentioned theses, were examined. The more detailed aims and objectives are given in the following.

1.2.2 Aims and objectives of the present study

As mentioned above, due to the missing proof of concept on an intensive production basis one major subject of the present thesis was to evaluate the DRAPS as a whole. Unfortunately, it was not possible to build three or more identical DRAPS. Therefore, the experimental setup occurred primarily in the hydroponic unit. That means that the used DRAPS had only one RAS but different hydroponic treatments. Nevertheless, the specified view to the hydroponic was absolutely necessary because in DRAPS the hydroponic unit is directly dependent on the fish rearing system and the adjustment unit. DRAPS has especially to compete with intensive single vegetable production in soilless culture. Thus, the plant performance in DRAPS was compared to plant performance by conventional hydroponics. As such, not only the quantity output of the system but also qualitative aspects of the produced vegetables were investigated. The latter was of great importance because very little is known about the effect of DRAPS, or aquaponics in general, on fruit quality parameters. The connection point between both cycles, the recirculating aquaculture system (RAS) and the hydroponic unit, was another focus of the present thesis. Therefore, the nutrient and oxygen concentrations were measured to investigate whether the components between the RAS and the hydroponic unit do have an impact. In this context, some problems were identified and the DRAPS modified during the study to address these issues. A first evaluation of the modifications was also done and is part of the present thesis.

Nile tilapia (*Oreochromis niloticus*) is one of the most utilised fish for aquaculture and is relatively robust. Probably due to this fact the few studies to DRAPS were just working with tilapia. As such, the present thesis was also used to examine if the concept of DRAPS is also useful to rear fish species other than Nile tilapia (under conventional practical conditions). Therefore, in the 2nd experiment (Table 1) or production cycle, the more famous aquaculture fish in Germany, the African catfish (*Clarias gariepinus*) were produced.

Due to the knowledge that fish waste water contains more soluble solids and microorganism than tap water, additionally, the oxygen concentrations of the nutrient solutions used in conventional hydroponics and in DRAPS were monitored. The aim was to evaluate if there were differences and if the differences can be traced back to the use of process water.

The results, found by Delaide et al. (2016) or Goddek (2017), did not confirm the results of the first experiment (Table 1) of the present thesis. Therefore, a further experiment was conducted (experiment 3, Table 1). To investigate the different statements; a small scaled experiment was conducted where only a hydroponic unit and not a whole DRAPS were used.

To evaluate how sustainable DRAPS are, the fresh water and fertiliser saving potential were investigated in comparison to that used by conventional hydroponics. Furthermore, it was calculated how much emissions can be saved by replacing nitrogen fertiliser with fish waste water.

In summary, the overall aim of the present study was a proof of concept of DRAPS as an intensive food production system with the main focus on the hydroponic unit. Based on this, the following specific objectives were formulated to meet the overall aim:

i) Determination of tomato and lettuce growth in terms of quantity and quality aspects under conditions of double recirculating systems in comparison to conventional hydroponic vegetable production.
ii) Examination of nutrient solution quality in terms of oxygen.
iii) Investigation of the nutrient concentration and the effect of the connection point between fish and plant unit.
iv) Evaluation of the fertiliser-saving potential and the environmental relief using DRAPS compared to conventional hydroponics.
v) Testing if DRAPS can be operated successfully with African catfish.
vi) Evaluation of the modification of the used DRAPS.*
vii) Direct comparison of nutrient solutions (based on fish waste water (DRAPS) and fresh water (conventional hydroponics)) to investigate the influence of the natural microorganism community on plant growth.

*Due to first research results a modification of the original constructed INAPRO research DRAPS were carried out.

1.3 Set-up of the thesis and short methodical overview

1.3.1 Overview

The scientific background of the present thesis is described in the 1st chapter, the general introduction and gives a brief introduction to the topic. A further part of the general introduction is the description of the aims and objectives of the present thesis (subsection 1.2.2). Due to recently published theses regarding double recirculating aquaponic systems, subsection 1.2.1 briefly describes what the present thesis differentiates from the other both.

The current state of knowledge is outlined in the second chapter. This chapter starts with a description of the impact of hydroponics on the environment (2.1) and a brief overview of aquaculture systems, especially recirculating aquaculture systems (2.2). It is followed by the subsection aquaponics which is again divided into two subsections. While the 1st subsection (2.3.1) deals with conventional low-tech, single recirculating aquaponic systems (SRAPS) and its bottlenecks, the second subsection (2.3.2) gives an overview about the current status of high-tech, double recirculating aquaponic systems (DRAPS).

The present thesis is a cumulative work. To meet the defined objectives three main experiments were conducted which are presented here in six different research papers or proceedings, respectively, separated in the chapters 3 to 8. For better understanding Table 1 gives a schematic overview of the chapters, the related experiments and the respective met objectives were already described in 1.2.2.

Table 1: Schematic overview of the chapters of the present thesis, the related experiments and the respective met objectives (1.2.2).

<table>
<tr><td colspan="3">Chapter 1
General introduction</td></tr>
<tr><td colspan="3">Chapter 2
State of knowledge</td></tr>
<tr><td>Experiment 1
2015
INAPRO[1] research DRAPS
Nile tilapia + tomato</td><td>Experiment 2
2016 + 2017
INAPRO research DRAPS
African catfish + tomato</td><td>Experiment 3
2016
Separated experiment; only with hydroponic unit[2]</td></tr>
<tr><td>Chapter 3
Advanced aquaponics: Evaluation of intensive tomato production in aquaponics vs. conventional hydroponics
(objectives i, iii)

Chapter 4
The potential of double recirculating aquaponic systems for intensive tomato production
(objectives i)

Chapter 5
Oxygen consumption in recirculating nutrient film technique in aquaponics
(objectives ii)</td><td>Chapter 6
(2016)
An innovative suction filter device reduces nitrogen loss in double recirculating aquaponic systems
(objectives i, iii, iv, v, vi)

Chapter 7
(2016 + 2017)
Prospects and challenges of double recirculating aquaponic systems (DRAPS) for intensive plant production
(objectives v, vi)</td><td>Chapter 8
Lettuce (Lactuca sativa, variety Salanova) production in decoupled aquaponic systems: Same yield and similar quality as in conventional hydroponic systems but drastically reduced greenhouse gas emissions by saving inorganic fertiliser
(objectives i, iii, iv, vii)</td></tr>
<tr><td colspan="3">Chapter 9
General discussion</td></tr>
<tr><td colspan="3">Chapter 10
General conclusion and recommendations</td></tr>
</table>

[1]EU project: Innovative aquaponics for professional application.

[2]Hydroponic unit at Humboldt-Universität zu Berlin, Albrecht Daniel Thaer – Institute of Agricultural and Horticultural Sciences and fish waste water (Nile tilapia) from Leibniz-Institute of Freshwater Ecology and Inland Fisheries, Berlin

DRAPS = double recirculating aquaponic system.

For the experiments 1 and 2 a new designed DRAPS – the INAPRO[1] research DRAPS – were used (experiment 1 and 2). For the INAPRO aquaponic system some parts of the prototype DRAPS developed by Kloas et al. (2015) were enhanced. A detailed system design and the main changing of the INAPRO research DRAPS are exactly described in chapter 3.

1.3.2 Experiment 1

In the 1st experiment, Nile tilapia and tomatoes were grown and it was mainly focused on the development and productivity of the tomato plants. This experiment was the content of the chapters 3, 4, and 5. Chapter 3 was mainly focused on a direct comparison of DRAPS and conventional hydroponics at the same electrical conductivity (EC) level of the nutrient solution for the plants (EC 1.8). Chapter 4 focuses on the evaluation of an increased EC level in aquaponics (EC 3.0).

Beside the plant growth, experiment 1 was dealing with the evaluation of the fruit quality in terms of secondary metabolites. Another focus of this experiment was on the evaluation of the nutrient concentration in the fish waste water. For completion, the saving potential of fresh water and fertiliser were calculated and compared to conventional hydroponics.

The 4th chapter includes more detailed evaluation of plant growth parameters but also evaluates the fruit quality in terms of the nutrient concentration.

Parallel to the mentioned investigation, the 1st experiment was also used to determine the oxygen concentration in the cultivation trenches and is also part of chapter 5. For evaluation, the oxygen consumption was calculated and it was investigated whether the nutrient solution itself (without plants) affected the oxygen consumption.

The 1st experiment was furthermore used to calculate the total biomass output and to calculate the area production ratio between aquaculture and hydroponic unit.

1.3.3 Experiment 2

In the 2nd experiment, Nile tilapia was replaced by African catfish and was the basis of the chapters 6 and 7. African catfish were used for the first time as combination partner in DRAPS. Therefore, the second experiment was done to evaluate if African catfish are usable for DRAPS. Therefore the plant

[1]EU project: Innovative aquaponics for professional application.

production in terms of fruit quantity was investigated. However, the focus in experiment 2 was less on plant performance than on the modification of the original INAPRO research DRAPS (in the following referred to as `original DRAPS´). The modification is described in both chapters and was absolutely necessary due to nitrogen loss in the original DRAPS. This in turn resulted in a fertiliser reduction far below its potential. Due to the high nitrogen fertiliser saving potential, the evaluation of the second experiment was further focused on the potential to reduce greenhouse gas emissions. As in the 1st experiment, the biomass output and the area production ratio between aquaculture and hydroponic was calculated.

After modification the nutrient concentration in the fish water (samples in the RAS) were investigated on 3 and 15 consecutive days, respectively. The second measurement (15 days) was done only in 2017 and is part of the chapter 7.

1.3.4 Experiment 3

The 3rd experiment (chapter 8) was detached from the INAPRO research DRAPS, and water from a running Nile tilapia RAS was used to grow lettuce (Table 2). This investigation was done to scrutinise the 1st and the 7th objective (1.2.2) of the present thesis. Interestingly, while the 1st experiment was carried out, a study was published where contrasting results were found (Delaide et al., 2016). The authors found significant and much higher yields (+39%) in "DRAPS" (they did not use a whole DRAPS system) compared to conventional hydroponics. Therefore, the 3rd experiment was conducted to investigate the possibly better growth of lettuce in DRAPS than in conventional hydroponics.[2] Additionally, the effect of the natural microorganism community in fish water on plant growth was investigated.

For experiment 3 exact nutrient solutions were prepared by hand. Therefore the results were also used to calculate the fertiliser saving potential. Additionally, the environmental relief due to reduced fertiliser supply was calculated.

1.3.5 General discussion and conclusion

In chapter 9 the results of the different experiments are discussed together. Additionally, some contributing results, which could not be pre-

[2]Addendum: Interestingly the results found by Delaide (2016) were recently confirmed by Goddek (2018), even if the yield increment in "DRAPS" was clearly less strong (see chapter 9.1.).

sented in the chapters before, were added. A few selected topics, which were estimated as particularly important, are discussed in chapter 9. This includes the evaluation of plant performance, the nutrient management, the optimal fish to plant ratio, and the environmental relief.

The thesis closes with a general conclusion and recommendations (chapter 10). The chapter starts with a brief reflection on the specific objectives listed in subsection 1.2.2. Afterward a summary of advantages and disadvantages is given using the SWOT analysis format. . To give a comprehensive overview not only the results of the present experiments, but also the experience gained during the four project years of INAPRO will be considered in this part. It is followed by a summary and evaluation of the improvements of the used DRAPS. Finally, the thesis will close with an outlook and recommendations for further research.

2

State of knowledge

2 State of knowledge

2.1 Effect of hydroponics on the environment

At a global scale, a large and increasing quantity of the daily consumed vegetables is produced in greenhouses. The total greenhouse vegetable area in the world amounts 497 815 ha (Hickman, 2018). A large proportion of its production is in soilless systems, also referred to as hydroponics, where plants are grown without using soil as rooting media (Savvas, 2003). To feed the plants, the irrigation water used in hydroponics is enriched with nutrients. Hydroponics benefits from independence from many soil related problems (structure, pathogens, salinity) (Savvas, 2003). This results in less necessary chemical soil disinfection (van Os et al., 1993), increased plant quality (Buchanan & Omaye, 2013), higher yields (Rouphael et al., 2004) and cost-effectiveness (Grafiadellis et al., 2000; Stanghellini et al., 2003) and increased water use efficiency compared to soil cultivation (Rouphael et al., 2004; Stanghellini et al., 2003). The latter is especially increased using recirculating hydroponic systems. Simultaneously, intensive productions systems like hydroponics can have a highly negative impact on the environment (Stanghellini et al., 2003). For example, high amounts of pesticide application (Stanghellini et al., 2003) can result in emissions which pollute the air (van Os et al., 1993) and the production of high amounts of different waste materials (Ruijs & Van Os, 1990; Stanghellini et al., 2003) such as plastic and rock wool. In addition, greenhouse horticulture is associated with fertiliser leaching into the environment (Stanghellini et al., 2003), especially in open systems. In hydroponics, nutrients are usually supplied in excess (Kläring, 2001; Le Bot et al., 2001) and a regular replacement of recirculated nutrient solution in hydroponics is necessary (Kumar & Cho, 2014). Fertiliser does not only affect the environment negatively by leaching, but also has significant impact during its manufacture. The production of nitrogen fertiliser is an especially energy-demanding process and high amounts of greenhouse gases (GHG) (CO_2, N_2O, NO, and NO_2) are emitted during its manufacture (Wood & Cowie, 2004). According to Zhang et al. (2013), each ton of manufactured nitrogen fertiliser emits 8.2 t CO_2-equivalent in China. This is about 4.3% of total Chinese GHG emissions, even if these high emissions are associated with coal mining and a lack of technological advancement in manufacturing factories in China. In this context, the impact of nitrogen fertiliser manufacturing on the environment becomes apparent and Torrellas et al. (2012) and Antón et al. (2004) stated that the emissions due to nitrogen mineral fertiliser production have a major impact on global warming, eutrophication and air acidification. Besides this, the production of vegetable needs fresh water. As previous mentioned, the water use efficiency (WUE; kg yield per m^3 water supply) is much higher in soilless systems than in open field production (Stanghellini et al., 2003). However, the WUE varies strongly (23–66 kg m^{-3}) in soilless systems, depending on climate conditions and growth system (Pardossi et al., 2004).

2.2 Recirculating aquaculture systems

Aquaculture is the fastest growing sector in animal food production. In 2004 the world aquaculture production was 31.1% of total fishery production, which has increased to 44.1% in 2014 (FAO, 2016b). Any further increasing fish demand will be met mainly by aquaculture. Around 62% of global food fish production is predicted to be supplied by aquaculture in 2030. As such it is projected that aquaculture will remain one of the fastest growing sectors for animal food production (FAO, 2016b; Msangi et al., 2013). As a whole, fish is very healthy for human diet due to its high-quality protein especially essential amino acids (Adeyeye, 2009), n-3 and n-6 polyunsaturated fatty acids (Steffens & Wirth, 2005), vitamins, especially D, A and B (Ostermeyer, 1999) and various minerals (Adeniyi et al., 2012). However, like other intensive animal production systems, aquaculture has a large impact on the environment – in particular, when the production systems are directly in touch with natural and open water bodies. As such, aquaculture systems can significantly reduce the surrounding water quality in terms of suspended solids, dissolved oxygen and nitrogen and phosphorous content (Boyd et al., 2005; Tovar et al., 2000). Some additional effects are increasing sedimentation rates around farms and beneath sea cages, changes in biodiversity around cages, escape of fishes and the endangerment of the natural fish population as well as the spread of diseases Boyd et al. (2005). Due to increasing environmental and regulatory constraints more sustainable fish productions systems are required. The most sustainable systems for intensive fish production are probably recirculating aquaculture systems (RAS; Figure 1; (Martins et al., 2010)).

Recirculating aquaculture systems (RAS) enable high productivity and simultaneously low water and land use requirements, with minimal ecological impact. These systems allow maintenance of optimal environmental conditions continuously for the respective reared fish species (Martins et al., 2010). Because the water in RAS is recirculating and (partly) reused (90 – 99%; (Blancheton et al., 2007)) it is necessary that the water is treated to remove waste products excreted by the fish. The necessary main facilities of RAS are shown in Figure 1.

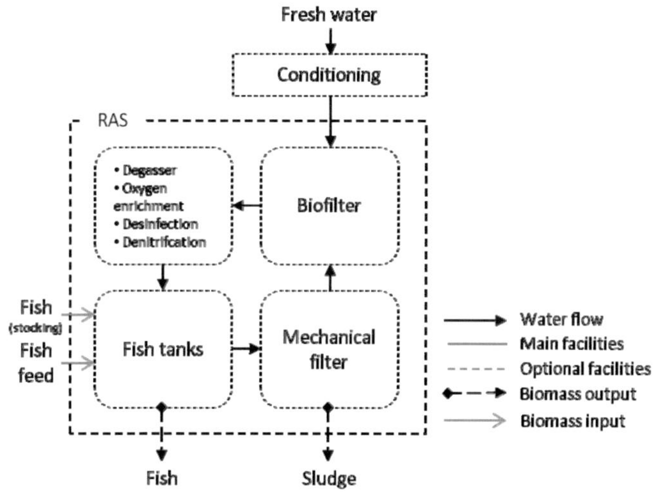

Figure 1: Schematic diagram of a recirculating aquaculture system (RAS) with its main components (fish tank, mechanical filter, and biofilter). Optional installations like CO_2 degasser (necessary when biofilter is a moving bed), oxygen enrichment, UV disinfection, and a denitrification unit can be added depending on requirements.

Briefly, the biofilter removes the fish-toxic ammonia/ammonium (see below). The mechanical filter removes the solid waste (sludge), composed of feed remnants, biofloc (dead and living bacteria), and faeces (Timmons et al., 2010). For mechanical cleaning, in general, three different removal mechanisms are differentiated: gravity separation, filtration, and flotation (Timmons et al., 2010). A detailed overview and classification of treatment functions are given by Vilbergsson et al. (2016). According to them, technical treatment functions are the sedimentation, mechanical filters (screens and granular media), floating bed bioclarifier, hydrocyclone, and flotation. Due to its importance in the present thesis, the sedimentation in settling basins and mechanical screen filter (drum filter) will be described briefly in more detail. Settling basins are one of the simplest technologies using gravity to separate the settleable solids due to the higher density of solid particles compared to water. For sedimentation, the water in horizontal or vertical containers is slowed down and particles allowed sink to the bottom (Timmons et al., 2010; Vilbergsson et al., 2016). The accumulated sludge has to removed immediately and regularly to prevent nutrient release and dissolution in smaller particles (Summerfelt, 1998). Rotating micro-screen drum filters with screen mesh pore size of 60 to 200 µm are the most frequently used in RAS (Badiola et al., 2012; Cripps & Bergheim, 2000). The operating principle is that a drum rotating on a horizontal axis, partially submerged into the water. Solid particles which are larger than the mesh size stick on the mesh. A backwash removes following the solids from the mesh by flushing it in a collecting tray (Timmons et al., 2010; Vilbergsson et al., 2016).

Due to their metabolisms fish excrete mainly nitrogen which is primarily in the form of ammonia (NH_3) (Evans et al., 2005). NH_3 is usually protonated immediately to ammonium (NH_4) after it enters the water (Timmons et al., 2010). Ammonia (NH_3) itself is highly toxic for many fish species (Ip et al., 2001) but its concentration present in the water depends among other things, on the pH value and the temperature of the water (Hobiger, 1996). With increasing pH and temperature the rate of NH_3 increases. Additionally, the pH determines not only the ratio of NH_3 but also the toxicity of NH_3 itself and increases with rising pH (Hobiger, 1996). Using the process of nitrification, the NH_4 in RAS is converted to nitrate (NO_3) by biofiltration. Nitrate is much less toxic to fish than ammonium/ammonia (Monsees et al., 2016; Timmons et al., 2010) but can rise in RAS to concentrations up to 500 mg nitrate nitrogen (NO_3-N) per litre (Honda et al., 1993). The NO_3-N concentration is crucial for the required daily water exchange rate (Timmons et al., 2010) in RAS. Besides nitrogen, other nutrients like phosphate, calcium, potassium, and magnesium can accumulate in RAS water, especially at low water exchange rates (Martins et al., 2009). In consequence, the effluent nutrient concentration which enters the environment is strongly increased (Blancheton et al., 2007). The problem is not only that nutrients like nitrogen and phosphorous reinforce eutrophication (Correll, 1998; Howarth & Marino, 2006) when they enter aquatic systems, but that nutrients, in general, are valuable resources. Therefore, the recycling of these nutrients would relieve environmental pressure and would make RAS more sustainable. One suitable approach to reuse nutrients from fish rearing is aquaponics where fish and plants are produced in a combined system.

2.3 Aquaponics

2.3.1 Low-tech, single recirculating aquaponic systems and its bottlenecks

The concept of synergetic production of fish in combination with other species is very old and was already applied by Incas and Aztecs. Nevertheless, the research on aquaponics "re-started" in the 70s and 80s of the last century. The pioneers on conventional single recirculating aquaponics are William McLarney (New Alchemy Institute), Mark McMurtry and Doug Sanders (North Carolina State University), and especially James Rakocy (University of the Virgin Islands) (Diver & Rinehart, 2010; Mageau et al., 2015; Thorarinsdottir, 2015). The primary idea of aquaponics is the use of synergetic effects. The nutrient-rich fish water is used for plant nutrition; where simultaneously the water is cleaned by plants and the bacteria living in the rooting zone (Diver & Rinehart, 2010; Rakocy et al., 2006). In conventional aquaponic systems, the fish unit and the plant unit are arranged in a single loop and referred to as single recirculating aquaponic systems (SRAPS; Figure 2).

The cleaning of fish water of solved agents depends on nitrification (conversion of ammonium to nitrate) by bacteria which settle in the rooting zone, and the nutrient uptake by the plants. A well functioning nitrification system is absolutely necessary in two respects. On the one hand nitrifying bacteria remove the fish toxic ammonium/ammonia from the water body and on the other hand nitrate is the preferred nitrogen form for many plant species (Hochmuth, 2001). In this context, plants usually grow best at combined ammonium and nitrate nutrition (Hawkesford et al., 2012), but many plant species are sensitive to ammonium when it exceeds a threshold (Britto & Kronzucker, 2002). For instance, tomatoes (*Solanum lycopersicum* L.) as species of the plant family of *Solanaceae*, are classified as ammonium sensitive plants (Britto & Kronzucker, 2002). As such, for hydroponically tomato production it is recommended that only 10% (Adams, 1999) to 50% (Claussen, 2002) of the total nitrogen concentration should be NH_4-N.

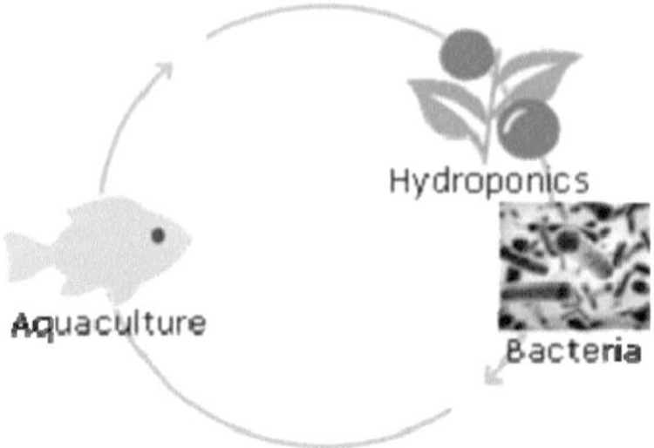

Figure 2: Simplified schematic diagram of single recirculating aquaponic system[3]

The essential components of modern SRAPS are fish tanks and a hydroponic unit. A mechanical filter unit, a biofilter for nitrification and a water storage can be also necessary but depends on the system design (Rakocy et al., 2006). The cultivation method in the hydroponic system is divergent. Plants are mainly cultivated in substrates like gravel, sand or perlite. Using substrates allows the elimination of a separate and expensive biofilter (Rakocy et al., 2006). However, some producers also use nutrient film technic (NFT) or vertical growing towers. A combination of different systems is also often used (Love et al., 2014; Rakocy et al., 2006). Aquaponic systems provide several advantages; for example the sharing of startup, operating and infrastructure cost. The double use of water and nutrients also reduces the environmental impact which makes aquaponics more sustainable. Aquaponic further facilitates also increased profit by producing two cash crops. Finally, less space is needed by the integrative production (Diver & Rinehart, 2010; Rakocy et al., 2006; Tyson et al., 2011). However, even if SRAPS shows great promise for sustainable food production there is a lack of information about factors of success and Tyson et al. (2011) suggest more long-term research is required. The difficulty is, that aquaponics itself is a highly complex system where three species, the fish, the plants and the nitrifying bacteria must be merged in one operating system (Tyson et al., 2011). But the combination of three different species causes challenges because each has different environmental requirements. In this context, the most critical environmental factors which have to be discussed are the pH value and the nutrient demand of the different species. Due to the essential removal of ammonia/ammonium from the water body to ensure fish wellbe-

[3]CC-BY-SA 4.0 EU-Projekt INAPRO, 2017

ing, the pH value in fish water depends on the optimum for nitrifying bacteria. While the optimum pH for most fish species range between 6.0 and 9.5 (Masser et al., 1992), the nitrifying bacteria show maximal specific growth rate at approximately pH 7.8 and increases with rising temperature (15 – 25 °C) (Antoniou et al., 1990). In fish water, the pH value can decrease due to the CO_2 production by fish and bacteria as well as from the acid produced by nitrification (ammonia-oxidation) (Bregnballe, 2010). In recirculating aquaculture systems, the pH value is usually stabilised by adding a base, like lime. To prevent ammonia stress for fish, it is suggested to level the pH value near the lower optimum pH value for nitrifying bacteria (Timmons et al., 2010; Tyson et al., 2004). In contrast, the recommended pH value in hydroponics is between 5.5 and 6.5 because the pH affects the availability of nutrients for the plants. PH values > 7 restrict mainly the availability of phosphate (Hochmuth, 2001; Trejo-Téllez & Gómez-Merino, 2012) while values above 8.3 also effect the availability of Mg and Ca (Trejo-Téllez & Gómez-Merino, 2012). At high pH levels, the micronutrients Fe, Mn, and Zn can become unavailable due to precipitation (Hochmuth, 2001; Resh, 2004). It was demonstrated that the growth of rose cultivars was negatively affected by a nutrient solution pH of 8 compared to control plants grown at pH 6 (Zieslin & Snir, 1989). Young cucumber (*Cucumis sativus*) plants showed a significantly reduced plant length, shoot fresh and dry weight at pH 8.0 when compared to pH 5.0 to 7.0. Furthermore, the early marketable cucumber yield was significantly decreased by pH 8 (compared to pH 5), however, the total yield was unaffected (Tyson et al., 2008).

The main source of nutrients in aquaponics is fish feed. On the one hand it delivers the nutrients for fish but it is on the other hand also the indirect source of plant nutrition. Lawson (1994) outlined in his paper water quality criteria for aquaculture. The criteria for selected nutrients are listed in Table 2. Additionally, published nutrient concentrations measured in RAS (Zechmeister, 2017) are listed and compared to the nutrient concentration in nutrient solution recommended for tomatoes by De Kreij et al. (1999), Hochmuth (2001) and Lattauschke (2004a), respectively.

Table 2: Comparison of recommended nutrient concentrations in nutrient solution for tomatoes, water quality criteria for aquaculture systems and measured nutrient concentrations in different aquaculture systems.

Nutrient	Recommended nutrient concentration for tomatoes[a] (mg L^{-1})	Water quality criteria for aquaculture[b] (mg L^{-1})	Nutrient concentrations in commercial operating intensive recirculating aquaculture systems[d] (min-max) (mg L^{-1})			Nutrient concentrations in SRAPS[e] min-max (mean) (mg L^{-1})
			Nile tilapia	Perch	African catfish	
NO_3-N	133–224.2	140/500[c]	56.1–170.3	46.6–67.2	106.2–181.2	1.9–110.5 (39.3^{12})
NH_4-N	7.0–16.8		0.3–1.9	0.02–0.19	47.3–116.4	0.03–2.9 (1.0^{6})
P	37.2–62	0.01–3.0	6.0–29.9	0.5–2.4	19.3–32.4	7.6–9 (8.2^{3})
K	199–371.5	<5	18.2–63.4	10.9–11.5	67.8–101.8	4.3–86.9 (40.3^{6})
Ca	128.3–216.5	4–160	28.4–48.0	41.8–48.5	110.1–114.0	5.1–34.2 (18.9^{6})
Mg	24.3–58.3	<15	13.0–21.0	10.9–11.5	25.4–32.1	1.7–8.6 (5.3^{5})
SO_4	144.2–446.9	50				3.8–24.4 (13.0^{3})
S	70	<1	39.4–67.0	36.0–39.3	51.2–73.4	
Fe	0.8–3	<0.15	0.01–0.06	0.01–0.04	0.09–0.12	0.21–3.0 (1.7^{3})
Mn	0.6–2	<0.01	0.02–0.06	0.003	0.15–0.23	0.02–0.1 (0.07^{4})
Mo	0.03–0.1		0.005	0.001	0.01	0.00–0.02 (0.01^{4})
Zn	0.09–0.4	<0.005	0.03–0.08	0.01–0.02	0.08–0.11	0.22–0.48 (0.38^{5})
B	0.3–1		0.05–0.08	0.04	0.07–0.10	0.05–0.23 (0.13^{4})
Cu	0.05–0.2		0.02–0.2	0.01	0.02–0.03	0.01–0.07 (0.04^{5})

[a]For production in rock wool, recirculating systems; references and recommendation are listed in more detail in Table 27

[b]By Lawson (1994).

[c]140 mg L^{-1} according to Schram et al. (2014) for African catfish 500 mg L^{-1} according to Monsees et al. (2016) for juvenile Nile tilapia.

[d]Samples collected from outflow of the fish tanks (N. tilapia, perch), or pump sump (A. catfish); N. tilapia = three, and perch = two fish tanks, at two dates (n = 6 for N. tilapia and n = 4 for perch) operating in Kirschau, Germany (Kirschauer Aquakulturen GmbH), stocking densities per tank = 16–33 kg m^{-3} (N. tilapia), 15–18 kg m^{-3} (perch); catfish = three RAS, one date (n = 3) operating in Lüchow, Germany, mean stocking density per RAS ≈ 240 kg m^{-3} (Zechmeister, 2017).

[e]SRAPS = single recirculating aquaponic systems without fertiliser addition; min and max values found in following references: Graber and Junge (2009), Palm et al. (2014b), Palm et al. (2014a), Rakocy et al. (2004b), Rakocy et al. (2004a), Roosta and Hamidpour (2011), Roosta and Hamidpour (2013), Roosta (2014), and Vergote and Vermeulen (2012); values in brackets are the mean values of all found min and max values in the respective references – the superscript number indicates the number of averaged values.

As shown in Table 2, the nutrient concentration measured in the different RAS differed from that recommended for tomato production in hydroponics (De Kreij et al., 1999; Hochmuth, 2001; Lattauschke, 2004a). While NO_3 and Ca is tolerated in a wide range by fish and can meet the demand of tomatoes, the nutrients phosphorus (P), potassium (K), magnesium (Mg), sulfur (S), iron (Fe), manganese (Mn) and zinc (Zn) is recommended in much lower concentrations for fish (Lawson, 1994) than for tomatoes produced in recirculating hydroponic systems. Less information about nutrient concentrations in fish waste water, discharged from commercial intensive recirculating systems is available in the literature. A few data could be found published by Snow et al. (2008), Ghaly et al. (2005) and Honda et al. (1993). Latter reported only data in regards to nitrogen concentrations. In terms of the first two publications, it was not clearly defined if the fish water was concentrated or from what kind of RAS the fish water came. For that reason, these data were not used in the following comparison. However, in 2017 water samples from a commercial Nile tilapia and perch operation were taken (Zechmeister, 2017) (Table 2) and will be used to characterise the nutrient concentrations of fish water in RAS. As shown in column 4 to 6 in Table 2, the recommended nitrogen concentration for tomatoes can be delivered by Nile tilapia and African catfish, but not by perch. Waste water from producing African catfish can contain enough Mg, S, and Zn (depending on the recipe) to support hydroponically grown tomatoes. In general, the fish water from African catfish contains the highest concentrations of nutrients compared to Nile tilapia and perch. But this might most likely due to higher stocking densities (240 kg m^3). However, while in RAS rearing African catfish can almost meet the recommended P and Ca concentrations, deficiency in K and most of the micronutrients is evident. In fish water gained from rearing Nile tilapia and perch the concentration of P, Ca, and S are additionally not sufficient to meet the demand of tomatoes.

In already operating SRAPS, the nutrient concentrations are usually below the recommended values for the plants in hydroponics (Graber & Junge, 2009; Palm et al., 2014a; Palm et al., 2014b; Rakocy et al., 2004a; Rakocy et al., 2004b; Roosta, 2014; Roosta & Hamidpour, 2011; Roosta & Hamidpour, 2013; Vergote & Vermeulen, 2012). In column 7 in Table 2 an extraction of published nutrient concentrations in SRAPS is shown. Only publications were chosen where no mineral fertiliser was added to increase the nutrient concentration. The low nutrient concentrations in SRAPS might be due to the usually low, or very low fish stocking density used in SRAPS, which can affect the nutrient content (Licamele, 2009; Pantanella et al., 2012). For example, Palm et al. (2014b) and Palm et al. (2014a) worked with final stocking densities (at harvesting) of 8.2 kg Nile tilapia per m^3 to 9.3 kg African catfish per m^3, respectively. In the study done by Roosta (2014) the final stocking densities were only 1.4 kg m^{-3} to 3.7 kg m^{-3} for different carp (*Cyprinus*) species. In commercial RAS Nile tilapia and African catfish are stocked at much higher densities. While the optimum level for Nile tilapia is around 80 kg m^{-3}, African catfish is stocked to >300 kg m^{-3} (Murray et al., 2014). Due to the lack of nutrients, some authors recommended the addition of specific nutrients to SRAPS to meet plant needs (Rakocy et al., 2004a; Rakocy et al., 2006; Savidov et al., 2005; Schmautz et al., 2017; Schmautz et al., 2016; Vergote & Vermeulen, 2012). As such, mainly Ca, K, and Fe (Rakocy et al., 2004a; Rakocy et al., 2006; Savidov et al., 2005) but also other macro (S, Mg, P) and micronutrients were added (Savidov et al., 2005; Schmautz et al., 2017; Schmautz et al., 2016). The comparison of the water criteria for aquaculture given by Lawson (1994) and the recommended values for tomatoes (Table 2) shows that P, K, Mg, SO_4, and micronutrients should be much lower for fish than for tomatoes. Consequently, an addition of nutrients to SRAPS should not be recommended.

For the combined production of Nile tilapia and lettuce (*Lactuca sativa*) SRAPS are already well established (Tyson et al., 2011). But also other plants like basil (*Ocimum basilicum*) (Rakocy et al., 2004b; Roosta, 2014), okra (*Abelmoschus esculentus*) (Rakocy et al., 2004a; Salam et al., 2013), water spinach (*Ipomoea aquatic*) (Salam et al., 2013), tomatoes (*Solanum lycopersicum*), cucumber (*Cucumis sativus*) and eggplant (*Solanum melongena*) (Graber & Junge, 2009; McMurtry et al., 1993; Roosta & Hamidpour, 2011; Schmautz et al., 2016) have been grown successfully in SRAPS. However, only a few studies are available where the production is compared to conventional single hydroponic production. Only the authors Pantanella et al. (2012) (lettuce), Roosta (2014) (basil), Graber and Junge (2009) (tomato, cucumber, and eggplant) have made such comparisons. The lettuce yield was dependent on the stocking density of the fish. In comparison to low-density aquaponics, the yield in hydroponics was higher. However, at high densities, the lettuce yield was similar (Pantanella et al., 2012). Basil performed much better in hydroponics than in aquaponics. The yield of tomatoes and cucumber was also better in hydroponics than in aquaponics while the yield of eggplant was similar between the both systems (Graber & Junge, 2009). The partly controversy results of the aforementioned studies were also confirmed by Wortman (2015). He conducted an experiment where he simulated aquaponics (low electric conductivity (EC) (0.5–1.0 dS m^{-1}) and high pH (7.2)) and conventional hydroponics (high EC (0.7–2.2 dS m^{-1}) and low pH (5.8)). While the relative growth rate of basil, kale (*Brassica oleracea*), hipotle pepper (*Capsicum annuum*) and cherry tomato did not differ between the treatments, the marketable yield for basil, kale, and tomato was significantly increased in hydroponics.

Finally, there are some well-established SRAPS. But due to the lack of information about factors of success and the merging of three different biological species most of aquaponic systems are only hobby systems, from non-profit organizations or for educative purposes and not for commercial production (Love et al., 2014). To overcome the problem of merging three different biological systems into a single synergistic system and to implement aquaponics as large-scale production units for intensive food production, the concept of aquaponics was rethought by Rennert et al. (2011) and Kloas et al. (2015). The result was the decoupling of the fish unit and the plant unit to produce both species more or less independent from each other while simultaneously maintaining the advantages of aquaponics. The key feature of such double recirculating aquaponic systems (DRAPS) is that the environmental conditions can be optimised for each species separately (Kloas et al., 2015).

2.3.2 Current status of high-tech, double recirculating aquaponic systems

The concept of double recirculating aquaponic systems (DRAPS), also referred to as decoupled aquaponics, is relatively new. The pioneers of DRAPS are probably Rennert et al. (2011) and Kloas et al. (2015). DRAPS consist of a separate, environmental friendly, recirculating aquaculture system (RAS) as described in subsection 2.2 for fish rearing and a recirculating hydroponic unit for plant production (Kloas et al., 2015; Rennert et al., 2011). The concept of DRAPS is shown as a simplified diagram in Figure 2. More detailed graphics are presented in chapter 3, 5, and 6.

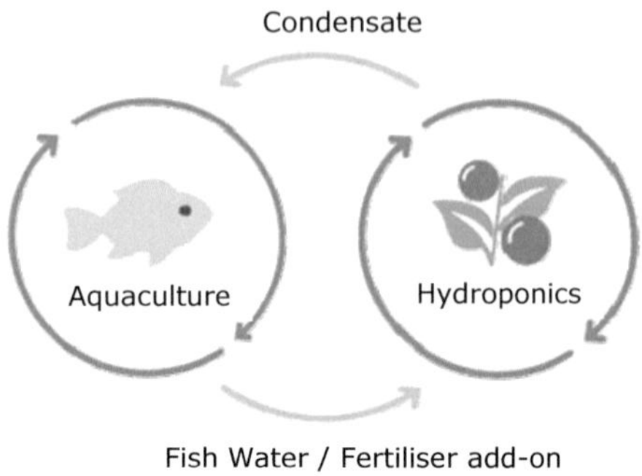

Figure 2: Simplified schematic diagram of double recirculating aquaponic systems[4]

DRAPS is based on the unidirectional water transfer from the fish to the plants. That means that the fish water is delivered from the RAS to the hydroponic unit, but not back from the hydroponic unit to the RAS, at least not directly (Rennert et al., 2011). This unidirectional water flow offers two main advantages. On the one hand fish can be reared independently from the nitrification performance and nutrient removal of the hydroponic unit. On the other hand, the systems enable the adjustment and optimization of the fish water in terms of nutrients (mineral fertiliser) and pH value for plant growth before it enters the hydroponic system. This, in turn, allows optimised intensive fish production as in conventional single RAS systems. In addition, the plant production can be specifically optimised for intensive production like in conventional single hydroponic systems (Kloas et al., 2015; Rennert et al., 2011). However, to maintain the idea of aquaponics as a closed system, Rennert et al. (2011) and Kloas et al. (2015) implemented cooling traps in the roof of the greenhouse. They were used to collect and regain the evapotranspiration water of the plants. This regained water was delivered back to the RAS, which reduces the overall fresh water consumption of DRAPS (Kloas et al., 2015).

A recently published review demonstrates that DRAPS becoming increasingly popular over SRAPS (Yep & Zheng, 2019). DRAPS or partly decoupled aquaponics have been increasingly mentioned in the literature since 2015 (Delaide et al., 2016; Goddek et al., 2015; Goddek & Vermeulen, 2018; Kloas et al., 2015; Monsees et al., 2017a; Monsees et al., 2017b; Thorarinsdottir, 2015). However, until now few empiric studies on DRAPS

[4]CC-BY-SA 4.0 EU-Projekt INAPRO, 2017

are available and the knowledge on DRAPS is low. As already mentioned, the first prototype of DRAPS as a complete system was built and investigated by Rennert et al. (2011) and Kloas et al. (2015). In the following years, some studies were conducted which were specifically related to the idea of DRAPS. As such, Goddek et al. (2016a) developed a theoretical design approach for a multi-loop aquaponic system including an anaerobic nutrient remineralization component to remobilise phosphorus. Karimanzira et al. (2016a) developed a dynamic model for DRAPS to simulate different aquaponic system scenarios, e.g. to define management strategies or to verify different system configurations. In a further work, Karimanzira et al. (2016b) applied optimal control methods to mathematical models. They showed how recirculating aquaponic systems can be operated more efficiently and economically. Further work on mathematic models for DRAPS was done by Reyes Lastiri et al. (2016). Their approach was to help understanding, analysing and optimising interacting water systems like DRAPS.

Besides the theoretical and model-based analyses, some investigations of single elements of DRAPS were conducted with the specific connection to DRAPS. As such, the potential of aquaculture sludge for nutrient remobilisation and reuse in the hydroponic unit was investigated (Goddek et al., 2016b; Monsees et al., 2017a). The mobilization of nutrients from fish sludge (feed leftovers, biofloc, and fish faeces; see 2.2) can proceed in an automated or semi-automated reactor (Monsees et al., 2017a). It is part of the multi-loop aquaponic system introduced by Goddek et al. (2016a) and enhances the sustainability of aquaponic systems due to optimised nutrient recycling (Monsees et al., 2017a). For example, fish sludge contains high amounts of phosphorus (6–30 mg g^{-1} TS), calcium (56 mg g^{-1} TS), iron (3 mg g^{-1} TS), and nitrogen (37–47 mg g^{-1} TS) (Lindell et al., 1996; Monsees et al., 2017a) which are important plant nutrients. While Goddek et al. (2016b) promoted an anaerobic nutrient mineralisation Monsees et al. (2017a) preferred an aerobic sludge treatment. Goddek et al. (2016b) investigated the effect of treated sludge on lettuce yield. In contrast, Monsees et al. (2017a) based his statement on nutrient analysis without examining the effect on plant growth empirically. Monsees et al. (2017a) demonstrated that aerobic sludge treatment mobilises phosphate and potassium and nitrate loss could be largely prevented. The amount of total ammonia nitrogen (TAN; ammonia (NH_3) + ammonium (NH_4^-)) and nitrite decreased to almost zero under aerated conditions. Contrary to this, during the anaerobic sludge treatment, no phosphorus was mobilised from sludge during the measuring period of 14 days. Additionally, the mobilization of potassium was even lower in the anaerobic than in the aerobic sludge treatment. Goddek et al. (2016b) assessed the sludge treatment according to lettuce growth in SRAPS. They found a significantly better growth when the supernatant from the anaerobic treatment was added to the nutrient solution (prepared with 85% tap water and 15% RAS water). They also measured the mineral nutrient concentration of the supernatant and interestingly, they did not find significant differences, except manganese. Manganese was however increased in the aerobic supernatant. Like in the study of Monsees et al. (2017a) the nitrogen form in the aerobic supernatant was mainly nitrate and in the anaerobic treatment was ammonium. Goddek et al. (2016b) suggested that the increased lettuce growth was a consequence of the higher ammonium concentration in the anaerobic supernatant.

The study of Monsees et al. (2016) focused on the chronic exposure of nitrate to juvenile Nile tilapia in RAS. Nutrients and especially nitrate can accumulate to high concentrations in RAS (Honda et al., 1993) and can become an issue in terms of fish health (Martins et al., 2010; Martins et al., 2009). However, the recommended NO_3-N concentrations for hydroponically grown plants vary from 100 mg L^{-1} to 240 mg L^{-1} for different crops (De Kreij et al., 1999; Hochmuth, 2001; Mattson & Peters, 2014; Snyder, 2016). Even if the information about critical nitrate levels are rare in literature, juvenile Nile tilapia tolerate NO_3-N concentrations up to 500 mg L^{-1} (Monsees et al., 2016). However, for African catfish only a maximum concentration of 140 mg NO_3-N L^{-1} is recommended to keep growth and feed intake at high levels (Schram et al., 2014).

Another study was focused on the growth of lettuce (*Lactuca sativa*) in DRAPS. Delaide et al. (2016) compared lettuce growth in DRAPS (tilapia water optimised and adjusted with mineral fertiliser), conventional low-tech aquaponics (pure tilapia water), and conventional hydroponics (fresh water). They found that the irrigation with optimised tilapia water increased the lettuce yield by 39% in comparison to conventional hydroponic or pure tilapia water. They assumed that substances like dissolved, organic matter with humic-like and protein-like components, or/and plant growth-promoting rhizobacteria and fungi contained in the fish water promoted plant growth. Recently, the experiment was repeated by Goddek and

Vermeulen (2018) in a slightly larger set-up. However, they detected only an 8% increase in fresh matter yield in supplemented aquaponics (equates DRAPS). Interestingly, Goddek and Vermeulen (2018) used 30% RAS water and added 70% rainwater without supporting reasoning. All the aforementioned studies above used just the idea of DRAPS. The experiments were not carried out in a full operating DRAPS but using fish waste water (adjusted with mineral fertiliser) for hydroponically grown vegetables like in experiment 3 in the present thesis.

The first investigation which considered a whole DRAPS was published by Kloas et al. (2015). They gave a detailed description of their self-developed DRAPS prototype. They demonstrated that DRPAS can be successfully used for the combined production of Nile tilapia (*Oreochromis niloticus*) at moderate stocking densities (38 – 45 kg m^{-3}) and tomatoes. In total, they produced 600 kg Nile tilapia (biomass gain) and 1 006 kg tomato fruit in around 265 production days. They calculated a production ratio of 1 kg fish feed plus 200 L fresh water input for 1 kg Nile tilapia and 5 kg tomato fruit output. The possibility to optimise the fish water for optimal plant growth was further demonstrated by Monsees et al. (2017b). They investigated the fish and plant production in DRAPS in comparison to SRAPS. While the fish production did not differ between both systems, the tomato fruit yield was increased by 36% in DRAPS even if they also added small amounts of mineral fertiliser to SRAPS.

Recently, Kledal and Thorarinsdottir (2018) reported a newly constructed double recirculating aquaponic system. Briefly, a 70 m^2 system with three 2 m^3 fish rearing tanks and three plant tables (1.45 m x 7.5 m) were built in Denmark at Institute of Global Food and Farming (IGFF; www.igff.dk; (Gutzmann & Kledal, 2014)). In South Iceland Svinna-verkfraedi Ltd. (www.svinna.is) constructed a pilot unit based on DRAPS with 12 m^3 fish tank volume and a 50 m^2 hydroponic unit. A commercial aquaponic system is developed by NER-Breen (www.breen.es) in Hondarribia, Basque Country, Spain. This system is only partly designed as DRAPS and its size is 6 000 m^2. The system is partly decoupled and the plant unit is mainly used to control the dissolved nitrogen level for the RAS (Kledal & Thorarinsdottir, 2018). Other commercial examples of DRAPS are driven by companies like Aqua4C (hortidaily.com, 2015) or ecf Farm Berlin (www.ecf-farm.de).

As has been shown, only a manageable amount is published about DRAPS and further research is necessary to make this sustainable approach competitive to single fish or vegetable production. Furthermore, it is noted that only Nile tilapia is used for evaluating DRAPS and validation with other fish species are non-existent.

3

Advanced aquaponics: Evaluation of intensive tomato production in aquaponics vs. conventional hydroponics

Johanna Suhl, Dennis Dannehl, Werner Kloas, Daniela Baganz, Sebastian Jobs, Günther Scheibe, and Uwe Schmidt

Agricultural Water Management 178 (2016) 335–34
DOI: 10.1016/j.agwat.2016.10.013

3 Advanced aquaponics: Evaluation of intensive tomato production in aquaponics vs. conventional hydroponics

3.1 Abstract

Aquaponics for intensive crop production is a highly complex system in which three different biological systems (fish, plants, and nitrifying bacteria) with different requirements must be merged. Finding the right combination is a serious challenge and the dependencies avoid a high productivity until now. Therefore, a unique and innovative double recirculating aquaponic system (DRAPS) was developed as a prerequisite for a high productivity comparable to professional stand-alone fish/plant facilities. It consists of two independent recirculating units – a recirculating aquaculture unit for fish production and a closed hydroponic cycle for plant production – which were connected unidirectional. This allows the use of fish waste water as nutrient supply for plants in hydroponics and its optimisation for plant growth by fertilizer supply without negative effects on fish rearing. Furthermore it allows a sustainable food production. In a new constructed DRAPS research facility, first investigations with tilapia and tomato production were conducted in 2015. During an annual production, it was demonstrated that in DRAPS comparable tomato yields were produced as obtained for conventional hydroponics. Even fruit parameters such as contents of lycopene and ß-carotene resulted in the same quantity when both systems were compared. Furthermore, the fertilizer use efficiency was increased by 23.6% in favour of the DRAPS. The total freshwater use efficiency was also increased using aquaponics.

3.2 Introduction

The world is confronted with the challenges of the 21st century including an increase in the world population, which is linked to a high demand of energy, food and water. These circumstances lead to climate changes, water and fossil fuel scarcities, soil degradation and food shortage. To face these challenges, sustainable food production with less water and energy consumption is becoming more and more important.

The Food and Agriculture Organization of the United Nations (FAO, 2014) reported that aquaculture is one of the fastest-growing food production sectors and provides already around 50% of all fish and fish products for human consumption. Additionally, a further intensification of aquaculture and an associated higher fish production is forecasted for the future. But intensive production systems are generally accompanied by drawbacks. Traditional aquaculture production in natural ponds or raceways causes significant negative environmental impacts, for example, the use of high amounts of fresh water and the hazard that a high nutrient load in the waste water can influence the environment negatively. Approaches to solve these drawbacks are recirculating aquaculture systems (RAS). These use 90–99% less fresh water than conventional systems and the nutrient release to the environment is accordingly reduced (Timmons et al., 2010; Verdegem, 2013).

Nevertheless, even RAS reduces the fresh water demand, a certain amount of fresh water is needed and the waste water is heavily accumulated by undesirable nutrients, especially of nitrogen. Aquaponics is an approach to minimise these negative impacts on the environment caused by RAS. Aquaponics consists of a combination of intensive fish production in pond culture and plant production in a hydroponic system (Klinger & Naylor, 2012). The additional production of plants may increase the added value and the sustainability of food production. The basic concept of this combined farming system is its use as a single recirculating aquaponic system (SRAPS), in which the waste water from fish production becomes available for plant production. The water is cleaned by the plants and bacteria living in the rooting zone and can afterwards be reused for fish production. This means that the fish wastewater flows into the hydroponic plant production system, passes the rooting zone of the plants and subsequently flows back into the fish rearing tanks. The technical composition of aquaponic systems, including a biofilter and/or a mechanical filter, can differ from system to system and is well documented by Diver and Rinehart (2010) and Rakocy et al. (2006). As such, aquaponics could relieve the environment by the double use of water and nutrients and increase the profit by producing two cash crops (Diver & Rinehart, 2010; Rakocy et al., 2006; Tyson et al., 2011). SRAPS have to struggle with different technical challenges (Goddek et al., 2015), because itis a highly complex system where three different biological systems (fish, plants, and nitrifying bacteria) must be merged into one working system. This difficulty prevents a high productivity, especially for fruit vegetables (Goddek et al., 2015; Vergote & Vermeulen, 2012; Wortman, 2015). In this context, one of the most critical points is the different pH optimum for fish, plants and nitrificants. While the pH optimum for fish and nitrifying bacteria ranged between 7 and 9 (Hochheimer & Wheaton, 1998; Rakocy et al., 2006), the recommended pH value for optimal nutrient availability in hydroponics varied between 5.5 and 6.5 (Hochmuth, 2001). If SRAPS are optimised, for example, for fish and bacteria (pH > 7), the availability of phosphorus, zinc, iron, manganese, copper and boron can be limited for plants (Hochmuth, 2001; Rakocy et al., 2006). For example, plants, such as rice, cassava, maize and French bean, are sensitive to high pH levels (>5.6–6.5) (Alam, 1981; Islam et al., 1980). Another crucial point arises from different nutrient requirements for fish and plants. The main input of nutrients in SRAPS is fish feed and consequently not only the nutrient source for the fish, but also, indirectly, for the plants as well. Nevertheless, a deficit of nutrients for plants is predictable when these would not be added to the hydroponic unit. Potassium, for example, has to be adjusted for plant production because the concentration released by the fish is not sufficient for plant growth. But also iron, calcium and phosphorus are usually insufficient in aquaponics and must be adjusted (Goddek et al., 2015; Rakocy et al., 2004a; Savidov et al., 2005).

SRAPS with combined production of tilapia and lettuce in raft systems seems to be well established (Tyson et al., 2011). Pantanella et al. (2012) have produced similar lettuce yields in aquaponics and hydroponics, but the yields were mainly dependent on the fish stocking density and nutrients were supplemented into the aquaponics. However, many growth parameters of mint and basil were negatively influenced by aquaponics (Roosta, 2014). Similar applies to tomato yields, which were significantly lower in aquaponics than those produced in hydroponics (Graber & Junge, 2009). Considering the latter investigation, very little is known about comparisons of intensive crop production in aquaponics and hydroponics under the same conditions (Nichols & Savidov, 2011). Tyson et al. (2011) suggested more long-term research, because the lack of information about factors of success is one reason why the most of aquaponic systems are only used as hobby or for education and not for commercial production (Love et al., 2014).

One approach to solve the mentioned problems caused by SRAPS was taken by Kloas et al. (2015). They have developed a unique and completely new double recirculating aquaponic system (DRAPS). The DRAPS addresses the challenge to achieve the food supply for the growing world population by intensive large-scale food production. To get an overview, the main advantages and disadvantages of SRAPS and DRAPS are listed in Table 3. One key advantage of DRAPS is the separation of the fish and plant cycle (Figure

3). As such, both production units are independent of each other. This means that the water quality, especially nutrients and the pH value, can be adapted to optimal conditions in both cycles separately. This improvement allows intensive production of fish and plants as in aquaculture and hydroponics, respectively. However, the construction, the fish water to plant ratio and the adjustment of the nutrient solution are not optimised (Kloas et al., 2015) and only few scientific studies exist about this technique.

Therefore, as a first step, the present study was focused on the optimisation of DRAPS. This include the installation of a 3-champer pit as part of the system, a spatial separation of fish and plant production, a reduced fish to plant ratio and the continuously adjustment of the nutrient solution applied for plants, in order to achieve equal amounts of tomatoes as generated in intensive crop production using hydroponics. The development and productivity of tomato plants grown under DRAPS conditions were compared with hydroponically produced ones.

Additionally, external and internal fruit quality parameters, such as blossom end rot fruit and contents of carotenoids, soluble solids and the sugar-acid ratio were analysed in tomatoes as well. These investigations were absolutely necessary because very little is known about fruit quality parameters caused by aquaponics. To demonstrate how sustainable DRAPS are working the fertilizer use and the total fresh water use efficiency were calculated and compared to that caused by conventional hydroponics. The optimal fish water to plant ratio was verified as well. Based on all results, first recommendations for successful intensive crop production using DRAPS are discussed.

Table 3: Advantages and disadvantages of single recirculating aquaponic systems (SRAPS) and double recirculating aquaponic systems (DRAPS).

Aquaponic system	Current advantages and disadvantages	References
SRAPS	**Advantages**	
	• combined production of lettuce and tilapia is well established	(Tyson et al., 2011)
	• well suited for hobby and education	(Love et al., 2014)
	Disadvantages	
	• suboptimal conditions for all integrated species, such as fish, bacteria and plants, in terms of pH, nutrients, water quality	(Goddek et al., 2015)
	• low production, especially on plant side	(Goddek et al., 2015)
	• too less scientifically research on direct comparison of aquaponics and single production of fish or plants	(Nichols & Savidov, 2011)
	• too little long term research	(Tyson et al., 2011)
	• not used for commercial production	(Love et al., 2014)
DRAPS	**Advantages**	
	• separation of fish and plant production enables optimal production conditions (pH, nutrients, water quality) for both, fish and plants	(Kloas et al., 2015)
	• optimisation measures in plant cycle does not affect fish cycle	(Kloas et al., 2015)
	• allows intensive crop production in large-scale	(Kloas et al., 2015)
	Disadvantages	
	• non-optimised construction, fish water to plant ratio, and adjustment of the nutrient solution	(Kloas et al., 2015)
	• too less scientific research	

3.3 Material and Methods

3.3.1 System design

The DRAPS used in the present study is based on the technology developed by Kloas et al. (2015). The experiments were carried out in a new constructed research aquaponic facility located in Abtshagen, Germany (52°31’12.025”N, 13°24’17.834”E). The total area was 196 m^2, which was divided into three areas: (i) technical room (14 m^2); (ii) the fish farm based on RAS (43 m^2); (iii) a Venlo-type greenhouse (139 m^2). The computer control system and a cogeneration unit were placed in the technical room. The RAS contained four identical glass fibre fish tanks with a total net production volume of 7.2 m^3. The water was cleaned by a mechanical filter (glass fibre sedimentation tank) with a volume of 1.3 m^3 and the effluent was collected in a pump sump with a volume of 2.34 m^3. From the pump sump the water was pumped to a trickling biofilter for nitrification to convert ammonium into nitrate. The specific surface area of the filter bodies was 120 m^2 m^{-3}. The nitrified water was collected in a reception water tank (0.4 m^3) and flowed back to the fish rearing tanks. The total volume of the whole RAS was around 12 m^3. Depending on the water quality and the fish stocking density, water treated with the mechanical filter was removed one to three times a week into the 3-chamber pit (3-cp; 4.5 m^3) (Figure 3).The fish water removing occurred unidirectional and discontinuously. From the 3-cp it was pumped into the storage tank (1 m^3) implemented in the greenhouse and was kept there until its use for hydroponically plant production (Figure 3). Before the fish wastewater was delivered to the plants, it was adjusted in the nutrient solution tank using mineral fertilizer to provide optimal nutrient concentrations for plant growth.

The greenhouse had a standing wall height of 4.2 m, which was equipped with double glass plates, whereas the roof was covered with single glass plates. The internal construction consisted of two single and two double gullies. Each double gully was restricted by a single gully, which was placed on the left or right outer wall, respectively. To evaluate plant characteristics, only the double gullies were considered. At the end of each gully, a nutrient solution tank with a volume of 300 L was installed. The cultivation method was based on nutrient film technique (NFT) applied in a recirculating system, in which the water was constantly pumped with a flowrate of 9 L min^{-1} for 24 h per day. The greenhouse was heated by a conventional floor level pipe heating system. Furthermore, a single layer energy screen and twelve high-pressure vapour sodium lamps were installed in the greenhouse.

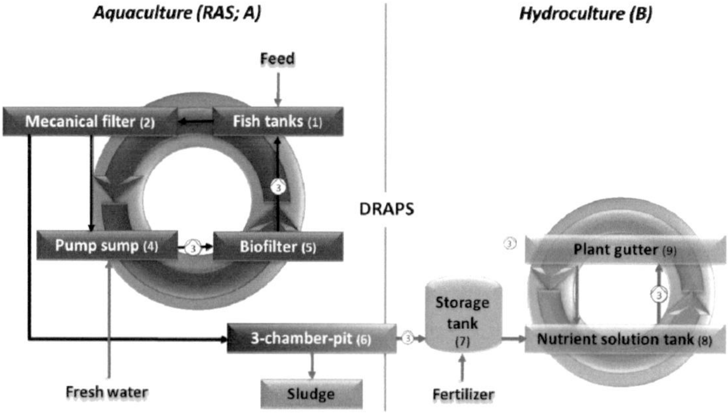

Figure 3: Schematic diagram of the used double recirculating aquaponic system (DRAPS). The recirculating aquaculture system (RAS; A): fish-rearing tanks (1), mechanical filter (sedimentation) (2), pump system (3) and trickling biofilter (5) and pump sump (4). Recirculating hydroponic unit (B): nutrient solution tanks (8) and plant gutters (9). Both systems are connected via a 3-chamber-pit (3-cp; 6).

3.3.2 *Investigational set up*

The experiments were conducted from 23.01.2015 until 08.01.2016. During this time, a combination of tilapia (*Oreochromis niloticus*) and tomatoes (*Solanum lycopersicum* L., cv. Pureza) were produced.

3.3.3 *Fish production*

The RAS was stocked with Red Natural Male Tilapia on 23.1.2015, 12.3.2015, and 15.7.2015. The initial stocking densities ranged between 2.0 and 2.3 kg m^{-3}. All batches were sorted once per growth period to separate bigger and smaller fish. The stocking densities during the whole fish production fluctuated between 2.0 kg m^{-3} and a maximum of 39.1 kg m^{-3}. During the investigations, the mean stocking density (20.7 kg m^{-3}) was kept low, because the fish batches were added and harvested stepwise. Other reasons for this practice were the activation of the biofilter, the fish feed, which changed in between to find an optimum, and some adjustments of the construction parts within the research plant during the growth period. All these factors mean stress for fish and led to suboptimal stocking densities. The feeding rate was between 0.7% and 2.4% and the daily water exchange rate in the RAS was 6.3% of the net fish production volume of 7.2 m^3. The whole fish production period was 351 days.

3.3.4 Plant growth conditions

During February until April, the set-points of the heating system were defined as 22 °C and 18 °C for day and night, respectively, and the ventilation was opened at 26 °C to cool the greenhouse. From mid of April until October the set-point for the heating system was changed to 17 °C for day and night and the set point for ventilation was dropped down to 21 °C. The energy screen was closed from one hour after sunset to one hour before sunrise and opened stepwise. The stepwise opening was controlled by a maximum temperature difference of two degree centigrade between roof and remaining greenhouse region. Between 7 a.m. and 8 p.m. the artificial light started at a level below 20 W m^{-2} global radiations.

For the experiments a net-acreage of 62.6 m^2 was planted with 144 tomato plants (2.3 plants per square metre). 48 plants per treatment were cultivated in two gullies. The plants were planted at 22.01.2015 when the first truss was visible. Two different treatments with two gullies each were applied. As such, one nutrient solution was mixed with fresh water and the other one with fish waste water referred to as hydroponics and aquaponics, respectively. The aquaponics treatment is shown schematically in Figure 3. The first application of fish waste water as process water to mix the nutrient solution was started at 05.02.2015. In this context, both nutrient solutions were adjusted with mineral fertilizer to an EC (electric conductivity) of 1.8 dS m^{-1} in order to achieve optimal growth conditions for tomatoes. To provide an optimal nutrient concentration, the nutrient content of both treatments was analysed periodically once a week. The target nutrient concentration was composed as follows: 151 mg L^{-1} nitrogen (N), 37 mg L^{-1} phosphorus (P), 234 mg L^{-1} potassium (K), 128 mg L^{-1} calcium (Ca), 24 mg L^{-1} magnesium (Mg), 110 mg L^{-1} sulphur (S), 2.0 mg L^{-1} iron (Fe), 0.3 mg L^{-1} boron (B), 0.2 mg L^{-1} copper (Cu), 1.2 mg L^{-1} manganese (Mn), 0.05 mg L^{-1} molybdenum (Mo) and 0.4 mg L^{-1} zinc(Zn) (Lattauschke, 2004a).

3.3.5 Determination of plant yield and plant growth

From 14.04.2015 until 27.10.2015 the tomato fruit were harvested once a week. To compare the total yield, the harvested tomatoes (n = 48) of each plant were weighed and counted. The fruit quality of ten plants per treatment (n = 10) were evaluated by categorizing the fruit into marketable fruit and non-marketable fruit. Latter were defined as fruit affected by blossom end rot (BER). The yield of both weight classes related to the total yield of the respective plant was used to calculate the yield percentage of both weight classes. Afterwards, the results were extrapolated on the canopy and expressed as kg cumulative yields per m2at the end of the experiments.

On four plants per treatment the number of leaves were counted and the leaf area (LA) per plant was calculated in three consecutive weeks in August 2015, in order to obtain results based on three biological repetitions (n = 12). For LA, the length and width of all leaves per plant (leaves > 20 cm) were measured with a folding rule. To calculate the leaf area, the data were integrated in an exponential function [3.1] developed by Dannehl et al. (2015a). At the end of the tomato production the total plant lengths of 6 plants per treatment were measured.

[3.1] Leaf area

$$LA = \frac{13{,}781 * e^{(0.075 * leaf\ length)} + 293{,}381 * e^{(0{,}027 * leafwidth)}}{2} - 164{,}809$$

3.3.6 Non-destructively method to determine leaf chlorophyll content

On four dates from June to August 2015, the remittance spectra of leaves were measured. As such, a hand held photodiode array spectrophotometer (Pigment Analyzer PA-1101, CP; Falkensee, Germany) was used to record the leaf spectra in the UV and visible wavelength range from 401 to 1048 nm (MMS1 UV–vis, Carl Zeiss; Jena, Germany). The measurement was carried out on three positions at the first fully developed leaf of each plant (n = 192). The remittance was evaluated at 750 (R_{750}) and 705 (R_{705}) nm to calculate the Chl NDI = Chl NDI = (R_{750} − R_{705})/(R_{750} + R_{705}) (Gitelson & Merzlyak, 1994; Richardson et al., 2002).

3.3.7 Sampling and preparation for chemical analyses of tomatoes

To determine the titratable acid (TA), the soluble solids content (SSC), as well as the content of carotenoids (ß-carotene and lycopene), three replicates per treatment containing five fruit each (>70 g) were randomly harvested. The sampling was repeated in three consecutive weeks in August 2015 in order to compare three biological repetitions (n = 9). The fruit were harvested at a ripening stage 9 (according to Organization for Economic Cooperation and Development (OECD) colour gauge) from trusses formed in the same height in the canopy. After harvesting, the fruit were frozen with liquid nitrogen and stored at −20 °C. Before chemical analyses were performed, the fruit were quartered. Two quarters were used to determine the dry matter (DM) content, whereas the rest of the fruit were mixed to a homogenate (KenwoodHB856, De'Longhi Deutschland GmbH; Neu-Isenburg, Germany).

3.3.7.1 Titratable acid, soluble solids and sugar-acid ratio

According to ASU-L-26.11.03-4 (1983) the TA in the homogenate was deter-mined by potentiometric titration. The exact procedure is described by Dannehl et al. (2014b).

For the determination of the SSC, 1.5 millilitre (mL) of the homogenate were filled into a 2 mL Eppendorf tube (EppendorfTubes®, 2 mL Eppendorf; Hamburg, Germany) and centrifuged at 6000 rpm for 5 min. After calibration with distilled (DI) water, the supernatant was measured using a digital refractometer (PR101, ATAGO; Karlsruhe, Germany). The refractometer detects reducing sugars and other soluble solids as °BRIX. One °BRIX is equivalent to 1 g soluble solids per 100 g fresh weight (g 100 g^{-1} FW). By dividing the content of SCC by the content of TA the sugar-acid ratio (SAR) was calculated.

3.3.7.2 Carotenoids

The extraction of lycopene and ß-carotene was conducted according to the method of Fish et al. (2002), which is based on hexane as solvent. The extraction and measurement is exactly described by Dannehl et al. (2014a). All measurements were carried out in duplicate and the content of carotenoids was calculated according to Nagata and Yamashita (1992). The results were expressed as milligram per g DM (mg g^{-1} DM).

3.3.7.3 Sampling and preparation for chemical analyses of nutrients in nutrient solutions

The samples to analyse the nutrients in the fish waste water and used nutrient solutions were taken from the storage tank (Figure 3, No. 7) and respective nutrient solution once a week (05.02.–20.10.2015) and stored in a freezer at −20 °C. After defrosting, the samples were filtrated and afterwards directly analysed using inductively coupled plasma-optical emission spectrometry (ICP-OES; P, K, Ca, Mg, S, and Na) and continuous flow analysis (CFA; ammonia nitrogen (NH_4-N) and nitrate nitrogen (NO_3-N)).

3.3.7.4 Nutrients in water samples

The nutrient analyses were conducted by ICP-OES (iCAP 6300Duo MFC, Fa. Thermo; Waltham, USA). Around 15 mL of the filtrated solution were used to flush the ICP. Afterwards, around 45 mL of the sample was used for measuring. The calibration curves for ICP-OES were established with the following reference solutions: blank 1.4 mol L^{-1} HNO_3; 0–100 mg L^{-1} of P; 0–300 mg L^{-1} of K;1–300 mg L^{-1} of Ca; 0–100 mg L^{-1} of Mg; 0–50 mg L^{-1} of S and 0–100 mg L^{-1} of Na. Additionally to the reported wavelength by Dannehl et al. (2015b) Na were analysed at wavelength 589.9 nm. The particular elements in the digestion solutions were analysed in duplicate and results were expressed as mg per litre (mg L^{-1}).

3.3.7.5 Nitrogen in water samples

The nitrogen content in the water samples were analysed using CFA (San++, Fa. SKALAR; Breda, Netherlands). The used reference solutions were 0.584 mg L^{-1} NH_4^+ (Ammonium standard solution, Merck KGaA; Darmstadt, Germany) and 0.565 mg L^{-1} NO_3^-- (Nitrate standard solution, Merck KGaA; Darmstadt, Germany). The standard series for CFA were established with dissolved NH_4NO_3 in the following concentrations: 0.25 mg L^{-1}, 0.5 mg L^{-1}, 1.0 mg L^{-1}, 2.0 mg L^{-1} and 3.0 mg L^{-1}. As rinse solution for the auto sampler, calcium chloride was used. As working solutions for NH_4-N analysis, one buffer solution (66 g potassium sodium tatrate + 48 g sodium citrate trihydrate + 6 mL Brij (30%) in 2 L DI water) and other solutions such as sodium salicylate (12.5 g NaOH + 40 g sodium salicylate in 0.5 L DI water), sodium nitroprusside (0.5 g in 0.5 L DI water) and sodium dichloroisocyanurate (1.0 g in 0.5 L DI water) were used. For analysis, NH_4-N was chlorinated to monochloramine. Latter react with salicylate to 5-aminosalycilate. After oxidation and coupling reaction, a green colour complex was formed. The adsorption of this colour complex was measured at a wavelength of 660 nm. For NO_3-N analysis, a buffer solution (3.405 g imidazole diluted in HCl + 3 mL Brij (30%) in 1 L DI water) and a staining reagent (50 mL H_3PO_4+ 5 g sulfanilamide + 0.25 g α-naphthylethylen-diamin-dihydrochloride in 0.5 L DI water) were blended. For analysis, NO_3-N was reduced to NO_2^- by a cadmium column. After NO_2^-reacted with sulfanilamide and coupled with α-naphthylethylendiamin-dihydrochloride, a red coloured azo dye was formed. This azo dye was colourimetrically measured at a wavelength of 540 nm.

NH_4-N and NO_3-N in the digestion solutions were analysed induplicate and results were expressed as mg L^{-1}.

3.3.8 Calculation of total fresh water and fertilizer use and its efficiencies

3.3.8.1 Water and fertilizer use

A water metre was used to measure the fresh water volume consumed by the RAS system during the whole fish production period. The same procedure was used to measure the volume of fish wastewater and the fresh water supplied to the respective plant treatment. Afterwards, the ratio from fish waste water to the fresh water supplied to the plants and the RAS, respectively, was calculated. The volume of the fresh water was further used to calculate the freshwater use efficiency of the control, described under Section 3.3.8.2.

To point out the fertilizer use, the amount of each nutrients added to the different treatments was weighed and summed until the end of the plant production. In these calculations, the loss of nutrients caused by regularly discarded volumes of nutrient solution used to prevent undesirable nutrient accumulation was not considered. The total fertilizer use is expressed in kilogram (kg). Fertilizer savings by aquaponics were calculated by using the amount of nutrient requirements in aquaponics in relation to that used in hydroponics, where the latter was given as 100%.

3.3.8.2 Fresh water and fertilizer use efficiency

The total fresh water use efficiency (FWUE) was calculated for both, the hydroponic and aquaponic system. In terms of the hydroponic system, the FWUE was calculated as ratio between marketable fruit and the total fresh water use. The results were expressed as kg tomatoes per m^3 fresh water (kg m^{-3}). However, the FWUE regarding the aquaponic system was calculated in the same manner, but under consideration of the total fresh water use for fish production in the RAS and fish waste water use for tomato production. As such, it was calculated how many kg fish and tomatoes can be produced with the same volume of fresh water.

The mean total fertilizer use efficiency (FUE) in terms of all nutrients for both treatments was calculated as a ratio between the total yield and the total fertilizer use (Brady et al., 2015) as described earlier. The results are expressed as kg total fruit yield per kg fertilizer supply (kg kg^{-1}).

3.3.9 *Statistical analysis*

The effect of nutrient solutions based on fish waste water and fresh water on plant parameters, such as yield, leaf area, number of leaves, Chl NDI, and on internal fruit quality parameters, such as contents of dry matter, SSC, SAR, lycopene and ß-carotene were analysed using SPSS package version 19.0. To compare the mentioned investigation parameters of both treatments, parametric t-tests were used after confirmed normal distribution by Kolmogorov-Smirnov test. Otherwise the non-parametric Mann-Whitney *U* Test was used. Mean values and standard deviations are displayed as numbers or bars in tables and figures, respectively. Statistical analyses were carried out on a significance level of $p < 0.05$, where significant differences between the treatments are characterised by small letters. Furthermore, the nutrients in the fish wastewater and different nutrient solutions were calculated by the same software program and in the same manner as described before.

3.4 Results

3.4.1 *Plant yield and plant growth*

The use of fish waste water with additional mineral fertilizer resulted in similar yields as in conventional hydroponic production (Figure 4). During the harvest period of 197 days, a total amount of 31.64 kg tomato fruit per m^2 was harvested in hydroponics, while in aquaponics was 29.38 kg m^{-2}, being not significant different. The rate of marketable fruit on total yield was nearly the same in both treatments (hydroponics = 99.1%, aquaponics = 99.5%). However, the production under aquaponic conditions resulted in a significant smaller rate (−0.37%) of fruit affected by blossom end rot (BER) compared to those ripened in hydroponics.

The plants grown under aquaponic conditions formed a significant smaller leaf area per plant (1.15 m^2) than those grown in hydroponics (1.36 m^2) (Table 4). However, the smaller leaf area in aquaponics was not accompanied by a lower number of leaves per plant. Furthermore the plant length did not show differences between the treatments (hydroponics = 10.8 m; aquaponics = 10.9 m). In contrast, the Chl NDI was increased by 17.9% in leaves developed by the influence of aquaponics compared to that in control leaves.

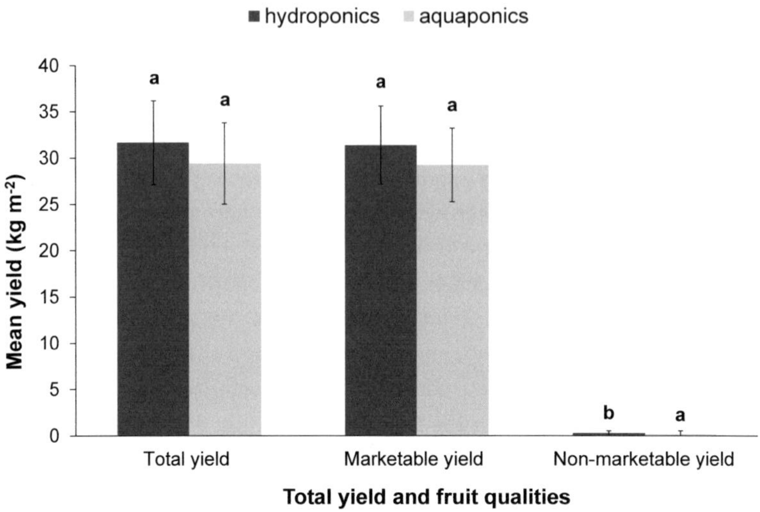

Figure 4: Effects of hydroponics and aquaponics on fruit yield and quality. The values represent the mean value of total tomato yield, marketable fruit yield and non-marketable fruit yield (n = 48) produced within 28 weeks ± standard deviation. The mean yields were tested using t-test and Mann-Whitney *U* Test, respectively and small letters indicate significant differences.

Table 4: Effects of hydroponics and aquaponics on leaf area (n = 12), number of leaves (n = 12), plant length (n = 6), and Chl NDI* (n =192) of the first fully developed tomato leaf.

Treatment	Leaf area per plant (m^2 $plant^{-1}$)	Number of leaves (Number $plant^{-1}$)	Plant length (m)	Chl NDI
hydroponics	1.36 ± 0.15^b	21.0 ± 0.7^a	10.8 ± 0.46^a	0.56 ± 0.19^a
aquaponics	1.15 ± 0.16^a	20.5 ± 1.3^a	10.9 ± 0.23^a	0.66 ± 0.16^b

The leaf area (LA), number of leaves (LN), plant length (PL) and remittance spectra indices Chl NDI content were compared using t-test (LA, PL) or Mann-Whitney *U*-Test (LN, Chl NDI), respectively. Data represent the mean of three (LA, LN) and four (Chl NDI) biological replicates ± standard deviation. Significant differences are indicated as small letters ($p < 0.05$).

*Chl NDI = Remittance (R) spectra indices for chlorophyll content.

3.4.2 Primary and secondary metabolites

The analyses showed a significantly lower DM content (−5.6%), SSC (−7%) and SAR (−5.0%) in tomatoes ripened under the influence of aquaponics compared to hydroponics (Figure 5). The carotenoid content in terms of lycopene (hydroponics = 0.99 mg g^{-1} DM, aquaponics = 0.90 mg g^{-1} DM) and ß-carotene (hydroponics = 0.20 mg g^{-1} DM, aquaponics = 0.19 mg g^{-1} DM) did not differ significantly between both treatments.

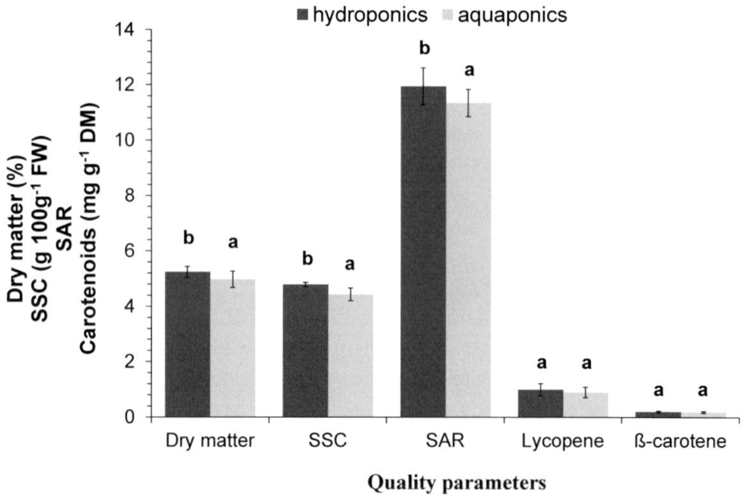

Figure 5: Influence of fresh water (control) and fish waste water based nutrient solutions on fruit dry matter content, soluble solids content (SSC), sugar-acid ratio (SAR), as well as lycopene and ß-carotene content. The data represent mean values and ± standard deviation (n = 9). The analysed contents were compared using t-test and represent the mean of three repetitions in three consecutive weeks. Small letters indicate significant differences ($p < 0.05$).

3.4.3 Nutrients in pure fish waste water, as well as in fresh and fish waste water based nutrient solutions

The amount of nutrients in the fish waste water used for plant production fluctuated strongly during the whole fish production period (Table 5). After passing a 3-chamber pit, its NH_4-N concentrations varied between 0.05 mg L^{-1} and 64.1 mg L^{-1} (mean 24.2 mg L^{-1}) and from unverifiable up to 42.7 mg L^{-1} (mean 14.6 mg L^{-1}) for NO_3-N. The concentrations of P, K, Ca, Mg, S and Na in the fish waste water fluctuated between 0.06 and 15.8 mg L^{-1},

3.2 and 69.1 mg L^{-1}, 54.2 and 119.8 mg L^{-1}, 9.2 and 19.3 mg L^{-1},15.3 and 50.4 mg L^{-1}, and 10.8 and 34.7 mg L^{-1}, respectively.

In a period of 8 weeks (02.07.2015 to 27.08.2015), the mean concentration of NH_4-N, K, Mg, and Na did not differ significantly in terms of the fresh water and fish waste water based nutrient solution (Table 5). However, differences in nutrient concentrations were observed regarding NO_3-N, P, Ca and S. NO_3-N, P and Ca were significantly lower (by 45.7 mg L^{-1}, 137.7 mg L^{-1} and 62.2 mg L^{-1}, respectively) owing to the nutrient solution mixed with freshwater. However, the S content was 30.7% higher in the fresh water based nutrient solution.

Table 5: Nutrients in pure fish waste water, as well as in nutrient solutions based on mixture of fresh and fish waste water, respectively.

Element	Pure* fish waste water		Nutrient solution based on** fresh water	fish waste water
	Mean ($mg\ L^{-1}$)	min. - max. ($mg\ L^{-1}$)	Mean ($mg\ L^{-1}$)	Mean ($mg\ L^{-1}$)
NH_4-N	24.2 ± 20.8	0.05 - 64.1	0.7 ± 0.9[a]	7.0 ± 9.0[a]
NO_3-N	14.6 ± 13.9	bld*** - 42.7	111.3 ± 19.4[a]	157.0 ± 50.3[b]
P	8.0 ± 5.0	0.06 - 15.8	67.2 ± 34.9[a]	204.9 ± 24.5[b]
K	30.2 ± 16.7	3.2 - 69.1	60.0 ± 34.9[a]	63.9 ± 43.5[a]
Ca	89.3 ± 20.1	54.2 - 119.8	165.0 ± 21.1[a]	227.2 ± 67.2[b]
Mg	13.9 ± 2.4	9.2 - 19.3	109.9 ± 27.7[a]	106.2 ± 28.5[a]
S	38.5 ± 8.9	15.3 - 50.4	216.8 ± 62.5[b]	150.3 ± 32.3[a]
Na	26.0 ± 5.3	10.8 - 34.7	130.1 ± 40.7[a]	126.9 ± 38.7[a]

*The data represent the mean content of nutrients in the fish waste water during the whole fish production period (23.1.2015 - 08.01.2016), including ± standard deviations and minimum (min.) and maximum (max.) values.
**The nutrient contents in fresh and fish waste water based nutrient solutions are given as mean values calculated for eight consecutive weeks from 02.07.2015 until 27.08.2015 ± standard deviation (n = 8). The detected concentrations were compared using t-tests. Significant differences are indicated by small letters ($p < 0.05$).
***blb = below limit of detection.

3.4.4 *Total fresh* water *use and fertilizer use and their efficiencies*

3.4.4.1 Fresh water consumption and total fresh water use efficiency

From 23.01.2015 (initial fish stocking) until the last day of tomato harvest (27.10.2015) the total fresh water consumption of the RAS amounted to be 160.4 m^3. The aquaponic treatment, with a total of 48 plants, consumed 13.6 m^3, i.e. 8.5% of the available fish waste water, from the day of planting (05.02.2015) until 27.10.2015. However, 14.2 m^3 fresh water were consumed by the hydroponic treatment during the plant growth period.

To investigate the FWUE, a mean yield of 13.05 kg $plant^{-1}$ in the aquaponic treatment was calculated. This resulted in a total fruit yield of 626.5 kg (Table 6), while the RAS achieved a total fish yield of only 248.84 kg. Considering these yields and the water consumption, it can be derived that in the present study 1.55 kg tilapia and 46.1 kg tomato fruit were produced with one m^3 fresh water applied to the RAS. However, the FWUE reached a value of 47.7 kg m^{-3} caused by the simple hydroponic system, when a total fruit yield of 677.3 kg and a fresh water consumption of 14.2 m^3 were considered.

Table 6: Total yield, total fertilizer addition, and fertilizer use efficiency (FUE) caused by hydroponics and aquaponics.

	hydroponics	**aquaponics**
Total yield per treatment (kg)	677.3	626.5
Mineral fertilizer addition (kg)	15.5 (100%)*	11.6 (74.8%)*
FUE (kg kg^{-1})**	43.7	54.0

*The percentages in the brackets represent the reduced fertilizer addition in aquaponics compared to hydroponics (100%).
**The FUE is calculated in kg produced tomatoes per kg mineral fertilizer addition.

3.4.4.2 Fertilizer use and fertilizer use efficiency

The addition of total fertilizer into the aquaponic system was reduced by 25.2% compared to that delivered to plants grown in a simple hydroponic system (Table 6). Based on the total yields and the total fertilizer uses as shown in Table 6, it was calculated that the FUE was improved by 23.6% in favour of the aquaponic system. This means that 10.3 kg more tomatoes can be produced with the application of one kilogram fertilizer into this system compared to a simple hydroponic system.

3.5 Discussion

3.5.1 Plant yield and plant growth

The goal to reach comparable yields with the innovative DRAPS as produced in the conventional way was achieved successfully. A total yield of 29.2 kg m^2 was gained using optimised fish waste water (Figure 4), and thus the new system can compete with modern hydroponic systems (Dannehl et al., 2014a; Dannehl et al., 2013). Kloas et al. (2015) achieved a total yield of 20.5 kg m^{-2} with the first prototype of DRAPS, where the same time duration of harvest period was considered. That means that the yield was increased by 1.4-fold caused by optimising the DRAPS, especially by the continuous measurement of nutrients and optimisation of the fish waste water for plant requirements (further discussed below) as shown in the present study. Roosta and Hamidpour (2011) demonstrated that in SRAPS equal tomato yields as in conventional hydroponics can be produced when different macro- and micronutrients were applied by foliar application. However, the drawback of the mentioned study is that the observation period was only 22 days and is therefore, not meaningful. Wortman (2015) found that the relative vegetative growth rates of basil, kale, tomato and pepper in simulated aquaponics (low EC + high pH) did not differ from plants grown in conventional hydroponics (high EC + low pH). But the marketable yields of all species were significantly reduced in aquaponics. He recommended a fertilizer supplementation to overcome these yield gaps. In the present study, it was demonstrated that with DRAPS comparable tomato yields as in conventional hydroponic can be produced. In addition, the rate of marketable fruit were slightly reduced in aquaponics (−0.5%) when compared with hydroponics (−0.9%) due to the fact that in hydroponics significantly more fruit were affected by BER (0.9%). The reduced amount of BER fruit in aquaponics (0.5%) might be caused by a higher Ca content in the fish waste water based nutrient solution (Table 5) (Millikan et al., 1971).

Although a significant reduction in LA (−15.4%) of plants grown in aquaponics was determined, the number of leaves and the yield were not clearly reduced simultaneously (Table 4). As such, the photosynthesis of plants is a function of the photosynthetic activity (source activity) and photosynthetic area (source size) (Engels et al., 2012). The lower assimilate supply by a reduced source strength in consequence of a reduced LA did not affect the yield obviously. One reason for this result might be the fact that the Chl NDI in leaves influenced by aquaponics was significantly increased (+17.9%). The Chl NDI is an indicator for the chlorophyll content in leaves (Richardson et al., 2002), where the chlorophyll content is directly linked to the source activity and therefore to the photosynthesis and consequently to the assimilate supply. Based on the higher Chl NDI value, it might be possible that the photosynthesis of plants grown under aquaponic conditions was more efficient followed by an accumulation of more primary metabolites in fruit, whereby the fruit yield was not influenced negatively.

3.5.2 *Fruit quality*

Plants grown in aquaponics developed fruit with a significantly lower DM content (5.12%) compared to those influenced by hydroponics (5.37%). However, in both treatments the DM content was in the range as reported for tomatoes (Frusciante et al., 2007; Herrmann, 1979). In the present study, the nitrate nitrogen content in the nutrient solution was significantly reduced in hydroponics (Table 5) during the ripening period. Bénard et al. (2009) reported that low nitrogen supply result in an increased DM content and consequently improved fruit quality in terms of SSC. Both, SSC and SAR were significantly increased in hydroponic production (+6.9%, +5.0%, respectively) (Figure 5). It has been reported that the overall flavour of tomato fruit is among other things depending on SSC and SAR (Baldwin et al., 1998; Stevens et al., 1979). Baldwin et al. (1998) found that SAR and SSC are positively correlated with an overall acceptability of tomato fruit among consumers. It was not proven, but it might be possible that the fruit flavour of tomatoes produced in hydroponics differed positively to tomatoes ripened under the influence of aquaponics. Due to the fact that the taste influence the purchasing behaviour of consumers (Pollard et al., 2002), the fruit produced by aquaponics are only competitive when they have a good taste. Therefore, a sensory analysis should be carried out in further experiments.

It is generally accepted that the diet and nutrition are central factors for a good health and that they play a key role in the prevention of chronic non-communicable diseases (NCDs) (WHO, 1990). NCDs are responsible for 63% of the global deaths (WHO, 2013). In this context, it has been reported that the consumption of tomatoes is associated with decreasing cardiovascular diseases and forms of cancer (Agarwal & Rao, 2000; Giovannucci, 1999). The main carotenoids of tomatoes are lycopene and ß-carotene, which are mostly responsible for these beneficial facts. They are antioxidants and react with free radicals or singlet oxygen (Di Mascio et al., 1989). In the human diet the major source of lycopene are tomatoes and tomato products (Stahl & Sies, 1996). Compared to results found for carotenoids in tomatoes by Frusciante et al. (2007), the lycopene and ß-carotene contents in fruit represented in the current study can be classified as high. It was reported that the lycopene content in tomatoes increased with increasing P or S supply, whereas this metabolite was negatively correlated to higher N and Ca supplies (Dumas et al., 2003; Montagu & Goh, 1990; Zelená et al., 2009). The same negative correlation applies between ß-carotene and different Ca applications (Dumas et al., 2003). In the present study, however, the fruit lycopene and ß-carotene contents did not differ significantly, although the content of S was significantly higher in hydroponics and that of NO_3-N, P and Ca content in aquaponics (Table 5) during the ripening period.

In terms of aquaponics, less information can be found about quality parameters regarding primary and secondary metabolites in plants. More research on quality parameters is necessary to provide not only sustainability facts about aquaponics, but also information about health-promoting plant compounds accumulated more or less in aquaponic products, which is necessary because quality aspects become more and more important for consumers (Pollard et al., 2002).

3.5.3 Water and fertilizer use and its efficiency

3.5.3.1 Water use and fresh water use efficiency

The plant unit consumed only 8.5% of the available fish wastewater (see 3.4.4.1). Extrapolated to the whole canopy, the plants in the greenhouse (144 plants) would consume 40.8 m^3, i.e. 25.3% of the fish waste water (from 161 m^3) generated by aquaculture and a daily exchange rate of 6.3%. This result suggests that the used research plant does not have the optimal RAS to greenhouse ratio. As shown above, approximately 161 m^3 fish waste water were available. Based on the results that 144 plants would consume 40.8 m^3 fish waste water, it can be calculated that 568 plants could be produced under consideration of 161 m^3 fish waste water, which is generated during fish production at a total productive volume of 7.2 m^3. With these numbers of plants and taking into account a plant density of 2.3 plants per square metre used in the present study, the optimal ratio between fish and plant unit would be a total productive volume of 7.2 m^3 and a net-acreage in the greenhouse of 247 m^2. Considering these data, a fish tank containing a volume of one cubic metre can be used to irrigate plants growing on a greenhouse net-acreage of 34.3 m^2. These calculations are strictly related to the combined production of tilapia and tomatoes as shown in the present study.

However, as already mentioned the set-up of the RAS for efficient tilapia production was hampered due to the changing environmental testing conditions during the experimental period. Therefore, productivity and stocking was much lower than described by Kloas et al. (2015). In the present study it was demonstrated that per cubic metre fresh water use only 1.55 kg fish but 46.1 kg tomato fruit were produced using the recently constructed DRAPS, whereas a not much higher yield was obtained (47.7 kg tomatoes) using the hydroponic system. Kloas et al. (2015) calculated that they could produce 5 kg tilapia and 25 kg tomatoes with the application of 1 m^3 fresh water to the RAS. The reason for the lower fish production per m^3 fresh water in the present study can be explained by two facts. First the much lower fish stocking density (20.7 kg m^{-3}) compared to that applied by Kloas et al. (2015). Secondly Kloas et al. (2015) used a technically higher advanced system to reuse the evapotranspirated water from plants and water surface via condensation in the RAS. The higher tomato yield production in the present study can be explained by a better control and fertilizer strategy in the present study compared to the study of (Kloas et al., 2015). In the present study the stock nutrient solution for the plants were adjusted regularly and very specific to the nutrient content of the used fish waste water. Furthermore, Kloas et al. (2015) stopped the continuous addition of micronutrients in between the harvest period. All this resulted in lower K and Fe contents in the nutrient solution for the plant unit than recommended for tomatoes in the mentioned study. These deficiencies might be a reason for the reduced yields. In this context it was reported that K has a positive effect on tomato fruit yield (Almeselmani et al., 2009). Roosta and Hamidpour (2011) found a positive effect of K and Fe foliar application on tomato fruit yield.

Finally, the research plant in the present study was built as a light version, i.e. without active cooling unit in the greenhouse. The installation of finned tube heat exchanger as an active cooling system in the aquaponic plant can lead to further water savings because it was found that 143 L m^{-2} condensation water can be collected during an annual tomato production (Dannehl et al., 2014c). A further advantage of this cooling system would be that the ventilation can be closed for a longer time resulting in higher CO_2 levels and an associated faster vegetative crop growth, higher yields and an increased accumulation of primary and secondary plant metabolites in fruit (Dannehl et al., 2014a; Dannehl et al., 2013). Furthermore, the water uptake of the plants can be reduced which results in a better water use efficiency (Dannehl et al., 2014b).

3.5.3.2 Fertilizer use and fertilizer use efficiency

The total fertilizer use in aquaponics was reduced by 25.2% compared to the control (Table 6). This was not as high as expected. The nutrient load, especially that of the nitrogen content in the fish waste water was unexpectedly low compared to the results by Kloas et al. (2015) using an established near-to-optimum productivity for tilapia. While Kloas et al. (2015) calculated a mean NO_3-N content of 127.7 mg L^{-1}, the used fish water in the present study contained only an average of 14.6 mg L^{-1}. The content of K, Ca, Mg and S was also reduced by 50.7%, 58.1%, 57.0% and 30.3%, respectively, compared to the results obtained in the study of Kloas et al. (2015). The lower nutrients in the present study were caused by the lower fish stocking density compared to that applied in the mentioned study. On the other hand, Kloas et al. (2015) achieved lower NH_4-N concentrations (5.79 mg L^{-1}) as found in the current study (24.2 mg L^{-1}). It might be possible that this circumstance was triggered by different measuring points. While Kloas et al. (2015) measured NH_4-N directly in the RAS before passing the fish water to the hydroponics reservoir tank, the present results were obtained by measurements in the storage tank after the fish waste water passed the 3-cp. Low ventilation of the 3-cp could be one reason for the higher NH_4-N content. Based on these results, it is assumed that an optimisation of the fish production in the RAS system and aeration of the 3-cp would deliver more nutrients for the plant production, which offers the possibility to increase markedly fertilizer savings in the range as described by Kloas et al. (2015).

The unique separation of the fish and plant cycle is one key advantage of DRAPS, due to the possibility to optimise the conditions (pH, nutrients) for fish and plants independently from each other. As shown in the present study, it is possible to realise also a good plant growth and fruit yields as in a simple hydroponic system with lower nutrient concentrations generated by the fish of the RAS, but only if sufficient nutrients are added to the nutrient solution for the plant unit. To ensure that nutrients are added in an optimal concentration for the plant growth, continuously measurements of the nutrients in the fish waste water and regularly adjustments of the fertilizers in the fish waste water based nutrient solution are indispensable. In this context, the best solution for DRAPS would be the installation of ion sensitive sensors with a direct feedback control as reported by Kläring (2001).

In the present study, it has been successfully demonstrated that the total yield produced in aquaponics is comparable to that in hydroponics and the fertilizer use efficiency was increased by approximately 23.6%. Applying 1 kg mineral fertilizer led to a fruit production of 54.0 kg in aquaponics, whereas only 43.7 kg tomatoes were produced in hydroponics (Table 6). However, if the fish productivity would have been in a similar range as demonstrated by Kloas et al. (2015), the fertilizer savings might have been much higher. This is the aim of ongoing studies.

3.6 Conclusion

It was shown that a double recirculating aquaponic system (DRAPS) with two independent cycles provides the opportunity to produce equal tomato yields compared to those obtained by conventionally used hydroponic systems. Moreover, it was demonstrated that even with a suboptimal fish production a good plant growth and fruit yield can be guaranteed using DRAPS. Nevertheless, to ensure that the added nutrients are in an optimal concentration for the plant growth, continuous measurements of the nutrients in the fish waste water and regular adjustments of the fertilizers in the nutrient solution are indispensable. DRAPS allows an intensive as well as a more sustainable food production. As such, it was found that the joint use of 1 m^3 freshwater resulted in 46.1 kg tomato production plus 1.5 kg tilapia, whereas only one food product, namely 47.7 kg tomatoes, was produced in hydroponics with the same volume of water. The fertilizer use efficiency was also improved by 23.6%. The results show that DRAPS contributes to lower the operating costs of plant production and to relief the environment, which is mainly based on the reuse of fish waste water and associated reduced quantity of nutrient emission. Additionally, the latter might be also interesting for fish farmer because in some countries the disposal is associated with high costs. Moreover, it was demonstrated that even the sustainable produced food with DRAPS has a high quality, similar to conventional produced products. The contents of lycopene and ß-carotene in fruit matured under present aquaponic conditions were comparable to those in the hydroponics ones.

4

The potential of double recirculating aquaponic systems for intensive tomato production

Johanna Suhl, Daniela Baganz, Werner Kloas, Dennis Dannehl, Sebastian Jobs, Günther Scheibe, and Uwe Schmidt

Acta Horticulturae 1242 (2019) 679–686
DOI: 10.17660/ActaHortic.2019.1242.100

4 The potential of double recirculating aquaponic systems for intensive tomato production

4.1 Abstract

Aquaponics combine the production of fish and plants in a sustainable way due to the double use of water and nutrients. Compared to conventional single recirculation aquaponic systems (SRAPS), the double recirculation aquaponic systems (DRAPS) are improved by the combination of two separate recirculating systems: the recirculating aquaculture system (RAS) for fish rearing and a closed hydroponic cycle for plant production. Both systems are connected in one direction to transfer the nutrient rich fish waste water to the hydroponic unit. The unidirectional connection of both systems allows the optimisation of fish waste water for plant growth by addition of mineral fertilizer without negative effects on fish rearing. DRAPS is a relative new approach and information about the functionality and successful use are rare. Therefore, more research on DRAPS is necessary. The present study was focused on different growth and production parameters of tomatoes produced in DRAPS compared to those produced in conventional hydroponics. Investigations during an annual production in 2015 showed that tomatoes produced in DRAPS have a comparable plant growth and development as obtained for conventional hydroponics. Even the fruit quantity and the fruit quality, including nutrient contents, were similar in both production systems.

4.2 Introduction

An aquaponic system is a food production technology that combines fish production in recirculating aquaculture systems (RAS) and plant production in soilless hydroponic culture. Aquaponics allow sustainable food production due to a double use of water and nutrients. Further benefits are a reduced environment pollution and an extra income by producing fish and plants (Diver & Rinehart, 2010; Rakocy et al., 2006; Tyson et al., 2011). Conventional aquaponic systems work as single recirculating systems (SRAPS), where the nutrient enriched fish water is used for plant production. During passing the root zone of the plants, the fish water in turn is cleaned by microorganism living in the rooting zone and by the plant itself due to the absorption of nutrients (Rakocy et al., 2006; Tyson et al., 2011). In the past, different aquaponic systems were developed and were equipped with different components. The optimum arrangement of SRAPS with their different existing systems are well-described by Rakocy et al. (2006) and Diver and Rinehart (2010). However, aquaponics consist of three different biological systems, including fish, plants and bacteria. These biological systems require different environmental conditions and must be merged in one operating system. Consequently, SRAPS struggle with different technical challenges, especially in consequence of different pH and nutrient requirements, which is well documented by Goddek et al. (2015). These challenges result in less productivity (Goddek et al., 2015; Vergote & Vermeulen, 2012; Wortman, 2015) and prevent the use of SRAPS as intensive crop production systems. Thus, most aquaponic systems are only used for hobby or for education (Love et al., 2014).

To overcome the mentioned problems of SRAPS and to implement aquaponics for intensive crop production, a double recirculating aquaponic system (DRAPS) was developed by Kloas et al. (2015). In DRAPS the fish and plant production is separated into two cycles which operate independently from each other. Thus, in both cycles optimal conditions such as the pH and the nutrient concentration can be adjusted separately. Consequently, intensive fish and plant production like in conventional single aquaculture and hydroponics can be realised (Kloas et al., 2015; Suhl et al., 2016). Nevertheless, the DRAPS approach is relatively new and less information of its functionality and its successful application is available.

To overcome this shortage, the present study was focused on tomato growth and development using DRAPS, where fish waste water from a recirculating aquaculture system (RAS) producing tilapia was used. It was investigated if the plant growth and development as well as the fruit yield and external and internal fruit qualities were comparable to those produced in conventional hydroponics. Therefore, the leaf area growth, generative development and fruit setting of tomatoes were compared during the first seven weeks within an annual production period. Additionally, the total plant length, the total number of trusses, as well as the cumulative fruit yield separated in marketable fruit and non-marketable fruit were investigated at the end of the production period. Furthermore, the concentration of some selected minerals in tomato fruit were analysed to determine the effect of DRAPS on fruit quality.

4.3 Materials and Methods

The experiments were carried out in a new constructed research DRAPS consisting of a 139 m² Venlo-type greenhouse, a 43 m² fish farm and a technical room with 14 m². For the experiment a combined production of red natural male tilapia (*Oreochromis niloticus*) and tomatoes (*Solanum lycopersicum* L., `Pureza´) were chosen. The experiment started with the delivering of the fish waste water to the greenhouse on February 5 and ended on October 27, 2015.

The fish waste water was collected in a storage tank placed in the greenhouse after it was passing a mechanical filter of the RAS and a 3-chamber pit. To provide optimal nutrient concentrations for tomato plants, the fish waste water was adjusted using mineral fertilizers. Therefore, the nutrient content of the fish waste water were analysed regularly. Three different treatments with two gullies each (n = 2) were applied: i) conventional nutrient solution, electric conductivity (EC) 1.8 dS m^{-1} (hydroponics); ii) fish water plus mineral fertilizer, EC 1.8 dS m^{-1} (aquaponics, EC 1.8); iii) fish water plus mineral fertilizer, EC 3.0 dS m^{-1} (aquaponics, EC 3.0). The target nutrient solution was prepared according to the mixture recipe of Lattauschke (2004a). The tomatoes grew in nutrient film technique and were planted with a density of 2.3 plants m^{-2}. For the experiment a total number of 144 plants were considered. Per treatment 48 plants were planted into two gullies (24 plants per replicate).

4.3.1 Plant growth and development

With starting of the application of fish waste water, five plants per treatment and replicate (n = 2 with 5 plants per replicate) were chosen randomly to count the number of leaves and to measure the total leaf area (LA) per plant. The measurements were carried out once a week in seven consecutive weeks. To calculate the LA, the length and width of all leaves (leaves > 20 cm) were measured with a folding ruler and the data were integrated in an exponential function [4.1] developed by Dannehl et al. (2015a) (Equation 3, Model no 6.). The results were expressed as cm^2 $plant^{-1}$.

[4.1] Leaf area

$$LA = \frac{13.781 * e^{(0.075 * leaf\ length)} + 293.381 * e^{(0.027 * leaf\ width)}}{2} - 164.809$$

To determine the generative development of the plants, the number of trusses (number per plant) of the mentioned five plants was counted during the first seven weeks of cultivation. Additionally, the fruit sets of six trusses (second to seventh truss) were investigated. The first truss was not considered because stress was caused by the delivery and planting resulted in a flower abortion. The fruit sets were calculated as percentage of harvested fruit per truss according to the number of formed single flower per truss. Results were expressed as mean value under consideration of the second truss onwards (%). Additionally, the total plant length in meter and the total number of trusses (number plant^{-1}) of three plants per treatment and replicate were measured and counted at the end of the experiment.

4.3.2 *Fruit yield*

During a period of 28 weeks, tomatoes from each treatment (n = 2 with 24 plants per replicate) were harvested once a week. The harvested tomatoes of each plant were weighted and counted. To compare the external fruit quality, fruit were further categorized into marketable and non-marketable fruit. As such, the latter were defined as fruit affected by blossom end rot (BER). In this context, only ten plants of the total plants (24) per treatment and replicate were used to determine the yield of both categorizations. Under consideration of these yields, the results were extrapolated on the total yield of 24 plants. The yields were cumulated and expressed as kg m^{-2} at the end of the harvest period.

4.3.3 *Mineral content of tomato fruit*

To determine the nutrient concentration in tomato fruit, 15 fruit per treatment, divided in three replicates (five per replicate), were harvested randomly. The fruit were harvested at a ripening stage nine (according to Organization for Economic Cooperation and Development (OECD) colour gauge) and at a single fruit weight of around 70 g. To compare three biological repetitions, the sampling was repeated in three consecutive weeks in August 2015 (n = 9). The mineral content were analysed by inductively coupled plasma-optical emission spectrometry (ICP-OES; iCAP 6300 Duo MFC, Fa. Thermo; Waltham, USA). For microwave digestion, 0.498 to 0.502 gram of freeze-dried material was weight into special, de-ionised vessels. The following procedure and the digestion program is exactly described by Dannehl et al. (2012). The calibration curves for ICP-OES were established with the following reference solutions: blank 1.4 mol L^{-1} HNO_3; 0 - 200 mg L^{-1} of K; 1 - 200 mg L^{-1} of Ca; 0.5 - 50 mg L^{-1} of Mg; 0.5 - 5 mg L^{-1} of Fe and 1 - 5 mg L^{-1} of Zn. The respective elements in the digestion solution were analysed at following wavelengths: K = 766.5 nm, Ca = 317.9 nm; Mg = 279.1 nm; Fe = 259.9 nm and Zn = 206.2 nm. The results are expressed as mg 100 g^{-1} FW.

4.3.4 Statistical analysis

The effects of different nutrient solutions on leaf area, leaf number, number of trusses, fruit sets, total plant length, total number of trusses and on fruit yield, as well as on mineral content of fruit were analysed using SPSS package version 19.0. For comparison of the mentioned parameters analysis of variance (ANOVA) were used and significant differences were calculated using Tukey-tests or Dunnett-T3 (variance heterogeneity) at a significance level of $p < 0.05$. In case of absence of normal distribution, the non-parametric Kruskal-Wallis-test with Bonferroni-test were used. Significant differences between the treatments were indicated using different small letters. The standard deviation was illustrated by ± in tables or bars in figures.

4.4 Results & Discussion

4.4.1 Plant growth and development

The use of fish waste water with additional mineral fertilizer did not affect the leaf area expansion within one week of comparison (Figure 6). The mean leaf area expansion from the first to the seventh week of production was 99.5, 99.2 and 112.3 cm^2 $plant^{-1}$ under hydroponic, EC 1.8, aquaponics, EC 1.8 and aquaponics, EC 3.0 conditions, respectively. The differences within one week were not significant. In contrast, Roosta and Hamidpour (2011) reported a reduced vegetative growth of tomatoes in SRAPS compared to hydroponics.

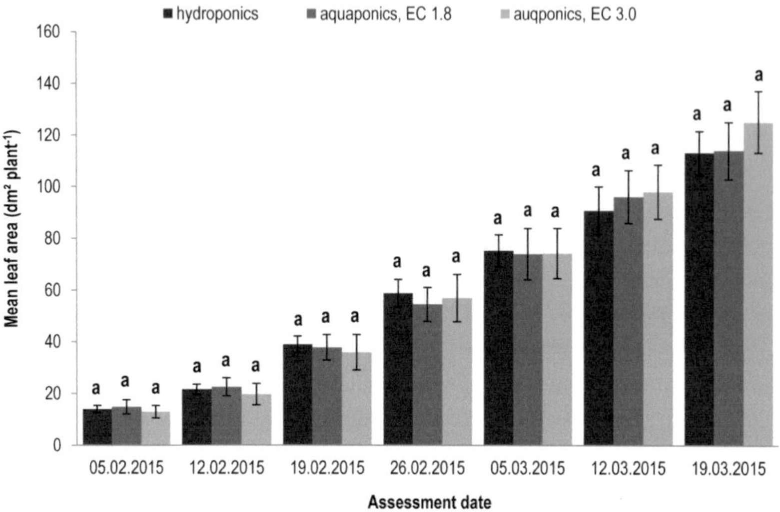

Figure 6: Effect of fresh water (hydroponics, EC 1.8), and two fish water based (aquaponics, EC 1.8 and EC 3.0) nutrient solutions on leaf area growth per plant in seven consecutive weeks. The mean leaf area was compared within one week using analysis of variance (ANOVA) and small letters indicate significant differences according to Tukey-test or Dunnett-T3 procedure.

In the present study, a similar developmental pattern was found in terms of number of leaves and the final length of the plants (Table 7). In contrast, Roosta and Hamidpour (2011) showed that the number of leaves per plant and the plant height was significantly reduced by SRAPS.

Furthermore, the generative development, including number of trusses during first seven weeks and total number at the end of the experiment, were also not affected in the present study. Even the fruit sets did not differ between the treatments.

Table 7: Influence of fresh water (hydroponics, EC 1.8) and fish water based nutrient solution (aquaponics) on number of leaves per plant, total plant length, number of trusses, fruit setting and total number of trusses.

Parameter (unit)	Week	Variant[1]		
		hydroponics	aquaponics, EC 1.8	aquaponics, EC 3.0
Number of leaves (no. plant^{-1})	1	7.6 ± 0.84[a]	8.1 ± 0.99[a]	7.7 ± 1.34[a]
	2	9.6 ± 0.52[a]	10.2 ± 1.40[a]	9.6 ± 1.51[a]
	3	12.0 ± 0.82[a]	12.5 ± 1.35[a]	11.9 ± 1.52[a]
	4	14.9 ± 0.99[a]	15.4 ± 1.43[a]	15.4 ± 1.78[a]
	5	17.2 ± 0.92[a]	18.2 ± 1.55[a]	17.7 ± 1.64[a]
	6	20.0 ± 0.94[a]	20.9 ± 1.66[a]	21.1 ± 1.91[a]
	7	23.3 ± 1.34[a]	23.5 ± 1.58[a]	23.7 ± 1.49[a]
Total plant length (m)		10.8 ± 0.46[a]	10.9 ± 0.23[a]	11.1 ± 0.08[a]
Number of trusses[2] (no. plant^{-1})	1	1.3 ± 0.46[a]	1.3 ± 0.64[a]	1.1 ± 0.30[a]
	2	1.9 ± 0.32[a]	1.9 ± 0.32[a]	2.0 ± 0.00[a]
	3	3.0 ± 0.00[a]	2.8 ± 0.42[a]	2.8 ± 0.42[a]
	4	3.9 ± 0.32[a]	3.7 ± 0.67[a]	4.1 ± 0.32[a]
	5	5.0 ± 0.47[a]	4.9 ± 0.32[a]	5.2 ± 0.63[a]
	6	5.5 ± 0.71[a]	5.5 ± 0.53[a]	5.3 ± 0.82[a]
	7	6.6 ± 0.52[a]	6.7 ± 0.67[a]	6.3 ± 0.67[a]
Fruit setting[3] (%)		93.5 ± 7.99[a]	90.9 ± 10.14[a]	93.2 ± 9.24[a]
Total number of trusses[4] (no. plant^{-1})		33.6 ± 1.52[a]	34.0 ± 1.0[a]	35.8 ± 0.50[a]

[1]Data represent the mean number of leaves, total plant length, number of trusses, fruit sets and total number of trusses ± standard deviation. Data were tested using Tukey-test and Kruskal-Wallis-test, respectively. Significant differences are indicated as small letters ($p \leq 0.05$).
[2]Number (no.) of trusses counted once per week.
[3]Mean fruit setting of the second to seventh truss.
[4]Total number of trusses counted at the end of the experiment.

4.4.2 Fruit yield

The total yield did not differ significantly between the hydroponic (31.64 kg m^{-2}) and the direct comaparable aquaponic variant with EC level 1.8 dS m^{-1} (-7.14%) (Figure 7). However, a higher addition of mineral fertilizer to the fish waste water (aquaponics, EC 3.0) increased the tomato yield significantly compared to aquaponics, EC 1.8 (33.08 kg m^{-2}). Nevertheless, the difference (+4.55%) with conventional hydroponics were not significant. The same applies to the marketable yield, while the rate of non-marketable fruit differed significantly among all different treatments. The smallest amount of fruit affected by BER was determined in aquaponics, EC 1.8 (0.18 kg m^{-2}), followed by hydroponics (0.29 kg m^{-2}) and aquaponics, EC 3.0 (0.42 kg m^{-2}). Despite a similar vegetative growth and generative development of the plants in the first seven weeks of the annual production period, the cumulative yield at the end of this period did not show the same results. One reson might be that plants in aquaponics, EC 3.0 formed heavier fruit than those grown under aquaponics, EC 1.8 (data not shown). However, the fruit produced under hydroponic conditions were also significant heavier than those in aquaponics, EC 1.8. Latter could have been deminished by a lower fruit number per plant in hydroponics (data not shown) and resultet therefore in comparable total yields even though the reduced fruit number was not significant. Anyway, the comparable yield in hydroponics and aquaponis, EC 1.8 confirmed the statement of Kloas et al. (2015) that tomato production in DRAPS can compete with tomoato production in single hydroponic production. But altogether, very little is known about tomato yields in aquaponics compared to those gained in conventional hydroponics. Moreover, the few steatments of similar yields produced in SRAPS compared to hydroponics (Graber & Junge, 2009; Roosta & Hamidpour, 2011) were based on very short-term experiments. Wortman (2015), hovewer, reported a significant reduced tomato yield in a simulated aquaponic system compared to hydroponics.

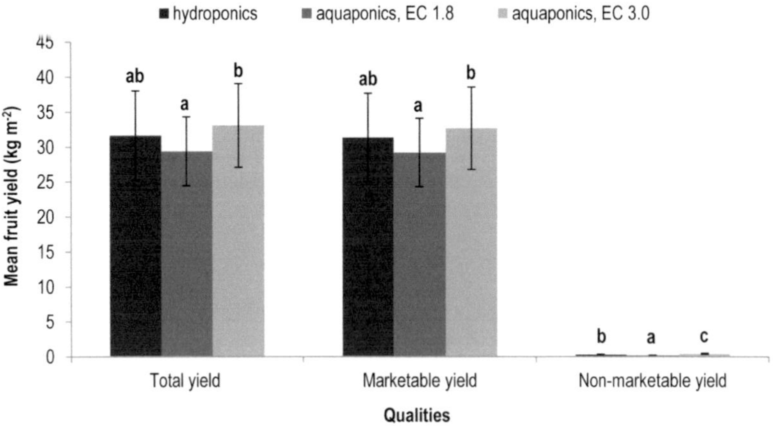

Figure 7: Influence of fresh water (hydroponics, EC 1.8) and two fish water based (aquaponics, EC 1.8 and EC 3.0) nutrient solutions on cumulative fruit yield. The data represent mean value of total tomato yield, marketable fruit yield and non-marketable (blossom end root affected fruit) fruit yield produced within 28 weeks ± standard deviation. The mean yields were tested using Tukey-test or Dunnett-T3-test and small letters indicate significant differences.

4.4.3 Mineral content of tomato fruit

From a nutritionally point of view, the fruit produced in conventional hydroponics had a better quality in terms of Mg (+10.2%) and Fe (+30.6%) compared to aquaponics, EC 1.8 (Table 8). The Zn concentration in turn was significantly increased caused by fish waste water based nutrient solution (both aquaponics +23.1%). The contents of K and Ca did not differ significantly between hydroponics and aquaponics, EC 1.8. The higher mineral fertilizer supply to aquaponics, EC 3.0 resulted in a significantly increased K concentration in tomato fruit. In contrast, Roosta and Hamidpour (2013) reported that tomato fruit produced in SRAPS reduced significantly the contents of K, Fe, Zn and Cu compared to those produced in hydroponics.

Table 8: Effect of hydroponics (EC 1.8) and aquaponics on nutrient content in tomato fruit.

Parameter	Element	Variant[1]		
		hydroponics	aquaponics, EC 1.8	aquaponics, EC 3.0
Mineral content in tomato fruit (mg 100 g^{-1} FW)	K	198.02 ± 9.11[a]	191.27 ± 16.12[a]	216.46 ± 8.48[b]
	Ca	9.58 ± 1.78[b]	7.93 ± 1.04[ab]	6.97 ± 0.84[a]
	Mg	7.41 ± 0.84[b]	6.65 ± 0.59[a]	7.39 ± 0.45[ab]
	Fe	0.36 ± 0.10[b]	0.25 ± 0.05[a]	0.28 ± 0.03[a]
	Zn	0.13 ± 0.02[a]	0.16 ± 0.02[b]	0.16 ± 0.01[b]

FW = fresh weight.
[1]Data represent the mean nutrient content of nutrients in tomato fruit (n = 9), ± standard deviation. Significant differences are indicated as small letters ($p \leq 0.05$).

4.5 Conclusion

The present study demonstrated that tomato production in DRAPS can compete with intensive tomato production in conventional hydroponics when using additional fertilizer. The unidirectional coupling of fish rearing and plant production enable an independent optimisation of both production units. Therefore, neither the plant growth during the first seven weeks, nor the cumulative fruit yield or fruit quality was influenced negatively when fish waste water-based nutrient solution was used. These positive results obtained in the present study confirm the big potential of DRAPS for intensive, but sustainable, food production.

5

Oxygen consumption in recirculating nutrient film technique in aquaponics

Johanna Suhl, Berry Oppedijk, Daniela Baganz, Werner Kloas, Uwe Schmidt and Bert van Duijn

Scientia Horticulturae, *255 (2019) 281–291*
DOI: 10.1016/j.scienta.2019.05.033

5 Oxygen consumption in recirculating nutrient film technique in aquaponics

5.1 Abstract

Due to root respiration, a minimum oxygen concentration in the root zone of plants is necessary. To keep the plant healthy and fully efficient the concentration should not fall below a plant specific critical value. Especially in nutrient film technique, the oxygen concentration can deplete heavily during the daytime. Besides the root respiration also microorganisms, living in the root zone of the plants, consume oxygen and have to be considered. In double recirculating aquaponic systems (DRAPS) fish waste water is used for preparation of nutrient solution for plant production in hydroponics. The fish waste water, delivered by a recirculating aquaculture system (RAS), probably contains high amounts of microorganism which can compete with plants for oxygen.

The present study was conducted to investigate the oxygen concentration of nutrient solutions used in both, conventional hydroponics and in DRAPS, respectively. Therefore, the oxygen concentration within the cultivation trenches of conventional hydroponics (control; nutrient solution prepared with fresh water, electrical conductivity (EC) 1.8 dS m^{-1}) and DRAPS (nutrient solution prepared with fish waste water (AP), EC 1.8 dS m^{-1} (low) and EC 3.0 dS m^{-1} (high)) were investigated. To evaluate whether the differences in oxygen depletion might be almost due to the used process water, the oxygen consumption of the prepared nutrient solutions was investigated separately without plants.

The oxygen concentration within the cultivation trenches followed a typical daily fluctuation pattern with a decreasing oxygen concentration during the day. As time passed from April to June the depletion during the day increased and was strongest in AP high. In June the oxygen concentration dropped partly to zero in AP high, while it dropped only to nearly 150 µmol L^{-1} and 60 µmol L^{-1} in control and AP low, respectively. The oxygen consumption of pure nutrient solution (without plants) was significantly different between AP high and control, which also affects the oxygen concentration within the cultivation trenches.

5.2 Introduction

Due to their energy demanding physiological activities (Morard & Silvestre, 1996) plant roots show a high respiration rate and therefore a high oxygen demand (Nielsen et al., 1998). According to Morard and Silvestre (1996) the root respiration depends on plant species and can differ from 1.44 µmol O_2 h^{-1} g^{-1} fresh matter (FM) to 7.8 µmol O_2 h^{-1} g^{-1} FM. The oxygen concentration in the nutrient solution can have an effect on plant performance in various forms (Bradford & Hsiao, 1982; Gislerød & Kempton, 1983; Morard et al., 2000; Morard & Silvestre, 1996; Stępniewski & Przywara, 1992) and is an important factor for plant growth. As such, a long term depletion of oxygen can lead to reduced plant growth and finally to declined yields (Gislerød & Kempton, 1983; Morard & Silvestre, 1996). A reason for growth depression can be a reduced water uptake by plants. It has been shown that the water uptake of tomatoes was reduced by 20%-30% after 48 h of oxygen depletion (Morard et al., 2000). This reduced uptake of water can cause stomatal closure and thus reduced transpiration (Bradford & Hsiao, 1982). However, not only the uptake of water but also the uptake of essential nutrients decreases due to oxygen deficiency (Stępniewski & Przywara, 1992). In terms of potassium, even an efflux from the roots into the environment was detected at low oxygen concentrations (Morard et al., 2000; Morard & Silvestre, 1996). Besides that, a good oxygen supply is also necessary for plant health. It was reported that a high oxygen concentration (11%-14%) prevented plants from infection with *Pythium* F707, while lower oxygen concentrations (5.8%-7% and 0.8%-1.5%) in the nutrient solution increased infection rate (Chérif et al., 1997).

In nutrient film technique (NFT) the roots are permanently surrounded by flowing nutrient solution, which appears as a barrier to gaseous exchange. However, in soilless culture systems oxygen deficiency mostly does not affect the complete root system, but specific parts are frequently affected. Even if only specific root parts are affected by oxygen depletion, it can cause a disturbance of the functioning of the root as described above (Morard & Silvestre, 1996). Gislerød and Kempton (1983) found in NFT trenches with cucumber an increasing depletion of oxygen from the inlet to the outlet. While the concentration was uncritical near the inlet (6.2 mg L^{-1}/193.8 µmol L^{-1}), it dropped to critical values for cucumber downstream at the last plant (2.9 mg L^{-1}/90.6 µmol L^{-1}). However, the depletion along the cultivation trenches was less when tomatoes were cultivated (decrease from 8.3 mg L^{-1} to 7.5 mg L^{-1}/259.4 µmol L^{-1} to 234.46 µmol L^{-1}). In contrast, the investigations of Holtman et al. (2009a) showed that the oxygen concentration in other NFT cultivation trenches for tomatoes can decrease to below 1 mg L^{-1} (31.3 µmol L^{-1}) during daytime in spring. This is below the critical value of 4 mg L^{-1} (≙ 125 µmol L^{-1}) for tomatoes (Zeroni et al., 1983). In deep water hydroponic systems, for example, oxygen supply to the plant root is the main problem (Zeroni et al., 1983).

The oxygen concentration in aqueous solutions, such as nutrient solutions used in horticulture, depends on different parameters. As such, the solubility of oxygen in aqueous solutions itself is affected. For instance, an increasing salt concentration (MacArthur, 1916) or a raising temperature (Carpenter, 1966) reduces the solubility of oxygen. In addition, oxygen concentration in nutrient solutions is affected also by biological factors. Plants withdraw oxygen due to respiration, and microorganisms in the nutrient solution and in the root zone consume oxygen as well (Jackson, 1980; Strayer, 1994). For conventional hydroponic systems, municipal/tap water, well water or ditch water, partly mixed with rain water, are used to prepare the nutrient solutions (Stanghellini & Kempkes, 2004). The amount of microorganisms and the number of species vary within the different water sources, and are usually low in well water (Strayer, 1994). For double recirculating aquaponic systems (DRAPS) fish waste water from recirculating aquaculture systems (RAS) is used to prepare the nutrient solution (Suhl et al., 2016). The RAS system contains a biofilter where microorganisms convert ammonium to nitrate (Tyson et al., 2011). Therefore, it can be expected that fish waste water from RAS contains much more and different microorganisms (Strayer, 1994; Sugita et al., 2005) than well water. Additionally, total suspended solids (TSS) accumulating in RAS and are a major source of carbonaceous oxygen demand (Timmons et al., 2010). This implies that the competition for oxygen between plants, microorganisms, and TSS is likely much higher in DRAPS than in conventional hydroponics, prepared with well water. DRAPS are developed for intensive fish and plant production (Kloas et al., 2015; Suhl et al., 2016). For intensive plant production optimal conditions, including oxygen concentration and gas exchange in the root zone, are essential and its assurance was the motivation for this study.

During an annual investigation of tomato production in a conventional hydroponics system and DRAPS the oxygen concentration in the nutrient solution was observed. The nutrient solutions of the different treatments were prepared with well water and mineral fertilizer for conventional hydroponic on the one hand and with fish waste water and mineral fertilizer for aquaponics on the other hand. To evaluate the different oxygen concentrations[5] within the nutrient solutions while passing the root zone, the oxygen consumption within the trenches was calculated. To get information about the impact of the plant activity a correlation analysis was done for the oxygen concentration within the trenches and the photosynthetically active radiation. The main focus of the present study was, to investigate if the nutrient solutions (prepared with well water or fish waste water) itself (without plants) differed in the oxygen concentration and to understand how strong this effect was. This information is necessary to understand hydroponics under DRAPS conditions and, if necessary to give recommendations to adjust and to improve the conditions for optimal plant growth, comparable to conventional hydroponic conditions.

5.3 Materials and Methods

The construction and design of the used double recirculating aquaponic system (DRAPS) is in detail described by Suhl et al. (2016) and pictured in Figure 8. The investigations took part in a Venlo-type greenhouse of a DRAPS (hydroponic, Figure 8) during the annual tomato production in 2015. The fish rearing (Nile tilapia; *Oreochromis niloticus*) is described in detail by Suhl et al. (2016). The hydroponic unit is described below, in paragraph 5.3.1. The recirculating aquaculture system (RAS) and the hydroponic unit are connected via a sedimentation unit, a storage tank and a preparing station to adjust the used fish waste water by adding mineral fertilizer and acid. The sedimentation unit consists of the mechanical sedimentation filter, as part of the RAS, and the 3-chamber-pit (3-cp). To perform the daily needed water exchange in the RAS, the fish water was released from the mechanical filter into the 3-cp to separate the sludge from the fish waste water. After the fish waste water passed the three chambers passively, it was pumped into a storage tank within the greenhouse where it was stored until its use as nutrient solution for plant irrigation. The sludge was separated within the 1st chamber of the 3-cp. The 2nd chamber was ventilated and the 3rd chamber was used as reservoir to pump the water into the storage tank.

[5] Taken from original publication; correct would be consumption

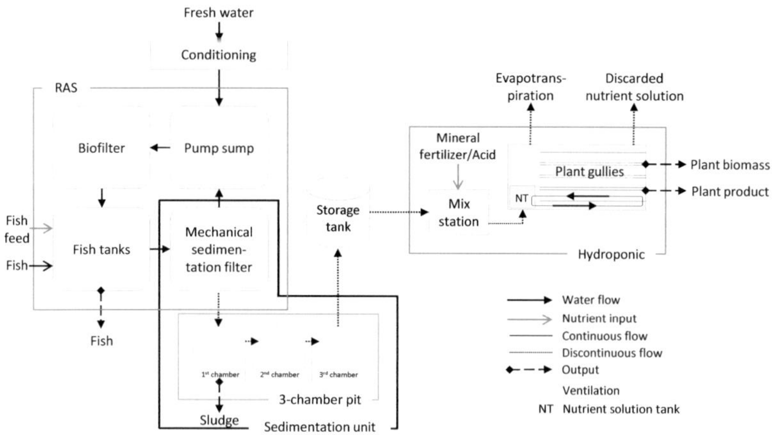

Figure 8: Schematic flow of the investigated double recirculating aquaponic system (DRAPS) according to Suhl et al. (2018b). The blue lines comprise the recirculating aquaculture system (RAS) and the green lines mark the hydroponic unit used for investigations. The RAS and hydroponic unit were connected via a sedimentation unit consisting of a mechanical sedimentation filter and a 3-chamber-pit, a storage tank, and the preparing station to adjust the fish waste water using mineral fertilizer and acid.

5.3.1 Plant production

From February until April 2015 the set-points for the heating system in the greenhouse were set to 22 °C and 18 °C (day, night). To cool the greenhouse the ventilation opened at 26 °C. From mid-April the set-points for the heating system were reduced to 17 °C (day, night) and the ventilation opened already at 21 °C. The energy screen opened stepwise one hour before sunrise and closed from one hour after sunset. The artificial light started at a level below 20 W m^{-2} global radiations between 7 a.m. and 8 p.m.

In the greenhouse two single trenches and three double trenches where installed (Figure 9) at a net-acreage of 82.7 m^2. The single trenches were placed at the left and right outer wall of the greenhouse, respectively, and were not considered for experiments. In total 192 tomato plants (*Solanum lycopersicum* L., cv. Pureza) were cultivated and grew in the recirculating nutrient film technique (NFT) system. In each 13 meters long trench a total number of 24 plants were planted. This resulted in a plant density of 2.3 plants m^{-2} in the greenhouse. The nutrient solution was constantly pumped from the nutrient solution tank to the end of each trench and circulated back in a free gradient (slope 1%) at a flow rate of 9 L min^{-1}. The plants were planted at 22nd January with about eleven leaves and when the 1st trusses were already visible.

Three different treatments with two trenches each were applied. Per treatment 48 plants were cultivated in two trenches. For the various treatments the nutrient solutions were prepared with different process waters. For the control treatment, the nutrient solutions were prepared with fresh water and for the two aquaponic treatments the nutrient solutions were prepared with fish waste water. To prepare the nutrient solution for the control, mineral fertilizer was added up to an electrical conductivity (EC) level of 1.8 dS m^{-1}. In case of the aquaponic treatments, the nutrient solutions were adjusted with mineral fertilizer to an EC level of 1.8 dS m^{-1} (hereinafter referred to as AP low) for the one treatment, and for the other treatment to an EC level of 3.0 dS m^{-1} (hereinafter referred to as AP high). The fish waste water was delivered from the associated recirculating aquaculture system, rearing Nile tilapia. For preparation of the nutrient solution an automatic mixing station controlled by EC and pH was used, as exactly described by Suhl et al. (2018b).

The use of fish waste water to prepare the nutrient solution started at 05.02.2015. To provide an optimal nutrient concentration in the nutrient solution, the fresh water for the control and the fish waste water for the aquaponic treatments were analysed periodically. The target nutrient concentration was composed according to Lattauschke (2004a) as following: 151 mg L^{-1} nitrogen (N), 37 mg L^{-1} phosphorus (P), 234 mg L^{-1} potassium (K), 128 mg L^{-1} calcium (Ca), 24 mg L^{-1} magnesium (Mg), 110 mg L^{-1} sulphur (S), 2.0 mg L^{-1} iron (Fe), 0.3 mg L^{-1} boron (B), 0.2 mg L^{-1} copper (Cu), 1.2 mg L^{-1} manganese (Mn), 0.05 mg L^{-1} molybdenum (Mo) and 0.4 mg L^{-1} zinc (Zn). The pH was adjusted to 5.8 using phosphoric acid. The corresponding EC level for this target nutrient concentration was EC 1.8 dS m^{-1} and this was applied to the control and AP low solutions. To reach EC 3.0 dS m^{-1} respectively higher concentrations were contained.

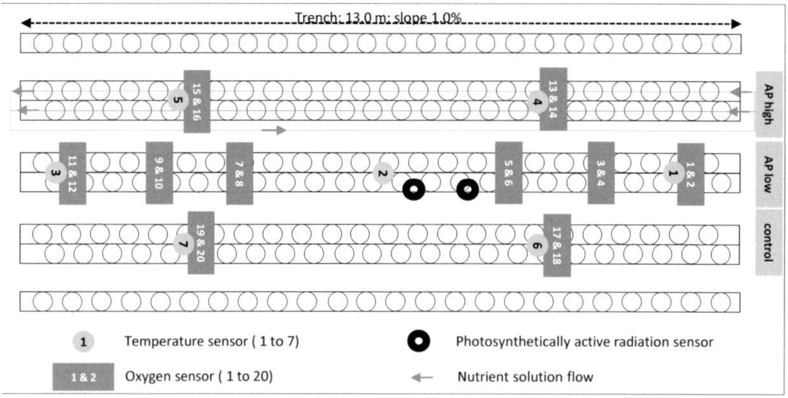

Figure 9: Ground plan and schematic indications of the control (electrical conductivity (EC) = 1.8 dS m^{-1}) and aquaponics (AP; AP low = EC 1.8 dS m^{-1}; AP high = EC 3.0 dS m^{-1}), as well as the locations of the temperature, oxygen, and photosynthetically active radiation (PAR) sensors.

5.3.2 *Measurement of photosynthetic photon flux density*

To measure the photosynthetic photon flux density (PPFD) in µmol m^{-2} s^{-1} two PAR (photosynthetic active radiation) sensors (type LI-190R Quantum; LI-COR, USA) were installed in the middle of the greenhouse (Figure 9). The sensors were installed just above the canopy of the tomato plants, and adjusted in height with the growing crop. The measurements occurred every 15 min and were saved automatically in a data base. For calculation the mean values of both sensors at each time point were used.

5.3.3 *Oxygen concentration/consumption measurements in the cultivation trenches*

5.3.3.1 *Oxygen measurement in the cultivation trenches*

To investigate the oxygen concentration in the nutrient solution passing the root zone of the tomato plants, a total number of 20 optical oxygen sensors were placed below the continuous flowing nutrient solution surface within the cultivation trenches. The oxygen sensors were installed in February 2015 and kept there during the whole plant production period until end of November. In one of the two trenches per treatment were four (control and AP high) or 12 (AP low) sensors installed (Figure 9), respectively. In each case two sensors were placed side by side. The distance between the sensor groups in the control and AP high treatments was 6 m. The respective first two sensors (13&14 and 17&18) were installed 3.5 m behind the inflow point of the nutrient solution. In AP low the first two sensors (1&2) were installed 0.5 m behind the inflow. The other sensors followed in a distance of 2 m, 4 m, 8 m, 10 m and 12 m from the first sensor.

For oxygen measurement plastic optical fibres were used as described by Holtman et al. (2009a). The fibres contained an oxygen sensitive fluorescent dye (Tris-Ru^{2+} 4,7 biphenyl 1,10 phenantroline) embedded in a polymer coating at the tip (Ast & Draaijer, 2014). All fibres were connected to a measuring device. Via blue light-emitting diodes (LED) blue light pulses with wavelength between 450 nm and 475 nm were sent through the plastic fibres. The life times of the emitted fluorescence light pulses from the dye, which increases with decreasing oxygen concentration, were detected by the measuring device. After digitizing, a processor calculated the oxygen concentration from the fluorescence life times and subsequently into percentage of saturated oxygen concentration (maximum 21%), based on calibrated relationships between oxygen and the fluorescence life time for each fibre. As the sensitivity of the oxygen sensors is slightly different for the different fibres a calibration procedure was applied to each fibre in which the oxygen concentration in air and in a 0% oxygen environment was measured for each fibre.

The oxygen concentrations in percentage in the nutrient solution were successively measured about every 15 min during day and night. Because not only the oxygen concentration itself, but also the fluorescence life time behaviour of the oxygen sensitive dye is affected by the temperature, a temperature correction was implemented for the dye. For this, additional seven temperature sensors were installed in the root zone to measure the temperature of the nutrient solution. Two sensors were used for the control and AP high and three for the trench were plants grew under AP low conditions. Except for one sensor in AP low trench (sensor 2, Figure 9) all temperature sensors were placed near the oxygen sensors. For correction, the slope and intercept of the correction curve was calculated. Due to its linearity, the actual oxygen concentrations ($O_{2,a}$) were calculated as in the following Eq. [5.1]. As such, $O_{2,m}$ means the measured oxygen concentration.

[5.1] Actual oxygen concentrations in %

$$O_{2,a}\ [\%] = slope * O_{2,m}\ [\%] + intercept$$

For representation in the present study, the actual oxygen concentrations ($O_{2,a}$) in % were converted into µmol L^{-1}. The conversion was performed in two steps. At first % was converted to mg L^{-1} (Eq. [5.2]). At this point, simultaneous a temperature correction occurred. In this case, $O_{2,sat,t}$ is the maximal oxygen concentration in aqueous solution at given temperature and $O_{2,amb}$ is the air oxygen concentration (21%). Using the molar mass of oxygen, the mg L^{-1} was recalculated into the µmol L^{-1} expression.

[5.2] Actual oxygen concentrations in mg L^{-1}

$$O_{2,a}\,[mg\ L^{-1}] = \frac{O_{2,a}\,[\%] * O_{2,sat,t}[mg\ L^{-1}]}{O_{2,amb}\,[\%]}$$

The data were displayed as average values of all corresponding oxygen sensors for each treatment. The chosen line diagrams illustrate the oxygen concentration in the course of the day at three different measuring periods. As such, the 1st measuring period was from 29.04.2015 until 05.05.2015, the 2nd period was from 31.05.2015 until 07.06.2015, and the 3rd period was from 11.06.2015 until 15.06.2015. Additionally, the temperature measured in the nutrient solution and the theoretical maximal oxygen concentration at a given temperature in pure water, were plotted.

5.3.3.2 Oxygen consumption in the cultivation trenches

To calculate the oxygen consumption ($O_{2,cons}$) in µmol L^{-1} h^{-1} the difference between the oxygen concentration (µmol L^{-1}) in the influx ($O_{2,i}$) and efflux ($O_{2,e}$) solution was multiplied by a factor which is the flow rate (Q) of 69 L h^{-1}[6], and divided by the volume (V; 64 L) of the nutrient solution in the trenches (see Eq. [5.3]). The time shift of around four minutes[7] the nutrient solution needed to flow from the inlet sensor to the outlet sensor was not considered because the concentration was measured only every 15 min.

[5.3] Oxygen consumption

$$O_{2,cons}\,[\mu mol\ L^{-1}\ h^{-1}] = \frac{(O_{2,i} - O_{2,e}\,(\infty))[\mu mol\ L^{-1}] * Q\ [L\ h^{-1}]}{V\ [L]}$$

The data were displayed as average values in total (day plus night) as well as for day and night separately. The data were averaged for two meas-

[6] Taken from original publication; correct would be 540 L h^{-1}
[7] Taken from original publication; correct would be seven

uring periods (marked as period A and B to keep it apart from the periods 1–3 used for oxygen measurement as described in 5.3.3.1). The period A was 01.05.2015 until 03.05.2015 and the period B was 05.06.2015 until 09.06.2015. Day and night were defined as time between sunrise and sunset for daytime and between sunset and sunrise for night time. For the calculation, the mean oxygen concentrations of the sensors 17&18 ($O_{2,i}$) and 19&20 ($O_{2,e}$) for control, sensors 1&2/3&4/5&6 ($O_{2,i}$) and 7&8/9&10/11&12 ($O_{2,e}$) for AP low, as well as 13&14 ($O_{2,i}$) and 15&16 ($O_{2,e}$) for AP high were used.

5.3.4 Oxygen consumption of the different nutrient solutions and fish (waste) water

5.3.4.1 Collection of samples and labelling

In a follow up experiment, the oxygen consumption of the nutrient solutions itself (without plants) was investigated (also referred to as jar experiment). The samples of the nutrient solutions were taken at two different points of the recirculating NFT system. The samples were taken on the one hand from the inflow of the nutrient solution into the cultivation trenches and on the other hand from the outflow of the trenches. Hereinafter these samples are labelled as "in" and "out" behind the relating treatment label, respectively.

Furthermore, the pure fish waste water used to prepare the respective nutrient solution for AP low and AP high was tested. As such, the samples were taken from the storage tank that was placed in the greenhouse (Figure 8). Hereinafter these samples are labelled as "fish waste water".

Additionally, the oxygen concentration in the fish water in the recirculating aquaculture system (RAS) was investigated. These samples were taken directly from the reception tank (between biofilter and fish tanks) of the RAS (Suhl et al., 2016) and are labelled as "RAS water".

For investigation, 6 samples from each variant were taken ($n = 6$), and a minimum of four of these were used to calculate mean values.

5.3.4.2 Investigation of the oxygen consumption of the different nutrient solutions and fish (waste) water

To investigate the oxygen consumption of the different waters (defined in 5.3.4.1), prepared glass jars with a total volume of 210.92 mL and air tight lids were used. The jars were prepared with a mount to fix the plastic optical fibres (Figure 10). The mounts itself were fixed with simple tape. For the measurement the same fibres and device as described in 5.3.3.1 were used, but here the oxygen sensitive fluorescent dye coating was not on the tip of the fibre but inside the glass jar at the position of the fibre mount. So, the contact point of the glass and the oxygen sensitive coating herein, was at the tip of the optical fibre outside the jar. This allows the measurement of oxygen concentrations in the airtight closed jar from the outside. For measuring, the glass jars were filled completely with the respective water samples without any air pockets left. Before filling the water into the jars, it was saturated with oxygen by shaking the solution vigorously in a large container. After closure of the jars, the oxygen fibres were fixed tight within the mount to ensure optical contact between the oxygen sensitive coating inside and the fibre outside the jar, and measurements started immediately. The oxygen measuring was performed as described in 5.3.3.1, but in this case the oxygen concentration was measured every five minutes for each jar. The measuring period per jar was restricted to a maximum of 370 min.

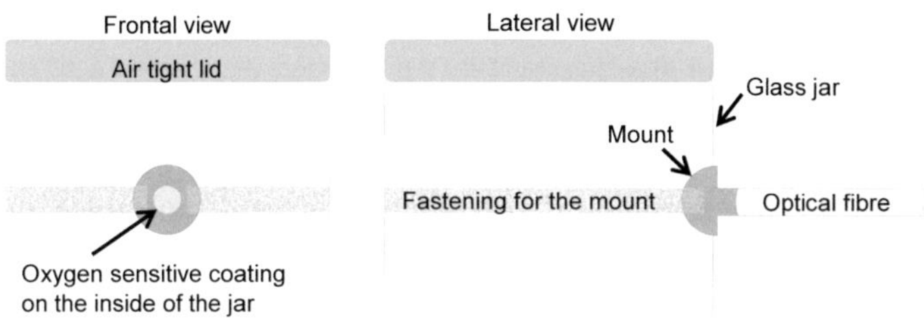

Figure 10: Schematic overview of the measuring jars used for oxygen consumption measurement.

For further calculations, the measured oxygen concentration in percentage was converted to µmol L^{-1} as described in section 5.3.3.1. To get information of the oxygen consumption rate ($O_{2,cons}$; µmol L^{-1} h^{-1}), the oxygen concentration in the water samples at the end of the measurement ($O_{2,t370}$) were firstly subtracted from the oxygen concentration at the beginning of the measurement (samples full saturated; ($O_{2,t0}$)). With the knowledge of the measuring time (370 min $\triangleq$ 6.22 h), the consumption per hour was calculated as shown in Eq. [5.4].

[5.4] Oxygen consumption rate

$$O_{2,cons}\ [\mu mol\ L^{-1}h^{-1}] = \frac{(O_{2,t0} - O_{2,t370})[\mu mol\ L^{-1}]}{6.22\ [h]}$$

Because of the different course of the decreasing oxygen concentration between the different waters, the oxygen consumption was partly calculated for different measuring periods. For all treatments the oxygen consumption for the whole measuring period (370 min) was calculated. For samples which showed a strong initial decrease, a threshold oxygen concentration was set. Above this threshold concentration of 173 µmol L^{-1} (period 1), the oxygen concentration decreased much faster, as below this concentration. For the samples which reached this threshold concentration, the oxygen consumption was calculated additionally for the time before and after 173 µmol L^{-1} was reached (period 2).

5.3.5 Statistical analysis

For statistical analysis the statistics software SPSS, package version 19.0 and 22.0, was used. The correlation between the oxygen concentration, the photosynthetic photon flux density (PPFD), and the temperature of the nutrient solution flowing within the cultivation trenches was determined using bivariate correlation. For evaluation, the Pearson correlation coefficients (r) were calculated and tested for significance levels of $p > 0.05$ and $p > 0.01$. The significance levels are characterised by * ($p = 0.05$) and ** ($p = 0.01$).

To analyse significant differences of the oxygen consumption within the cultivation trenches of the different treatments, and between the different nutrient solutions as well as the fish (waste) water, univariate analysis of variance (ANOVA) was used, after normal distribution was confirmed by the Shapiro-Wilk test. The pairwise comparison between the different variants was according to Tukey-HSD (homoscedasticity) or Dunnett-T3 (heteroscedasticity) tests. Not normal distributed data was tested using the Kruskal-Wallis test and pairwise compared using the Dunn-Bonferroni test. The comparison of two samples (oxygen consumption in cultivation trenches during the day and night) was tested using the Mann-Whitney *U* test (normal distribution failed).

Mean values and standard deviation are displayed as numbers in tables and statistical analysis was carried out on a significance level of $p < 0.05$. Significant differences between the treatments are indicated by small or capital letters.

5.4 Results

5.4.1 Oxygen concentrations in the cultivation trenches

The oxygen concentrations in the nutrient solutions were measured and investigated continuously, and were investigated. Figure 11A, 12B and 12C show in more detail the results for three measuring periods. Generally, for all measuring periods a typical daily fluctuation of the oxygen concentration was observed. While the oxygen concentration decreased during the day, it increased again during night time. In general, throughout the whole measuring period, from end of April until mid of June (Figure 11A, 12B, and 12C), an intensification of the oxygen depletion during daytime was detected. Especially the hot summer season in mid of June (3rd measuring period, 11.06. until 15.06.2015; Figure 11C) had a strong effect on the oxygen concentration. While in the 1st measuring period (29.04. until 05.05.2015; Figure 11A) the oxygen concentration during daytime dropped to a minimum of around 150 µmol L^{-1} for all treatments, the oxygen concentration dropped even partly to zero during daytime in the following weeks (3rd measuring period, Figure 11C). In this context, the correlation analysis showed that the correlation between oxygen concentration and temperature was negative and ranged between -0.607 (1st period) to -0.817 (3rd period) and is shown in Table 9.

However, in the 2nd measuring period it became apparent that the oxygen concentrations under different treatments became more different from each other. In the 2nd measuring period (31.05. until 07.06.2015) in both AP treatments a minimum of about 50 µmol L^{-1} (AP high) to 100 µmol L^{-1} (AP low) were detected, while the oxygen concentration in the control declined only to a minimum of 150 µmol L^{-1}. Primarily the AP high treatment showed a strong decreasing oxygen concentration, as it dropped towards 0 µmol L^{-1} during daytime in the 3rd measuring period (11.06. until 15.06.2015). In the same measuring period, the oxygen level in the control and the AP low treatments declined to a minimum of around 148 µmol L^{-1} and 57 µmol L^{-1}, respectively. In the night, the oxygen concentration under all three treatments levelled at around 250 µmol L^{-1} in the first two measuring periods. In the 3rd measuring period the oxygen concertation raised up to a maximum of around 250 µmol L^{-1}, 225 µmol L^{-1} and 200 µmol L^{-1} in the control, AP low, and AP high treatment, respectively. In summary, it was detected that the oxygen concentration in the AP high treatments was generally lower than in both other treatments.

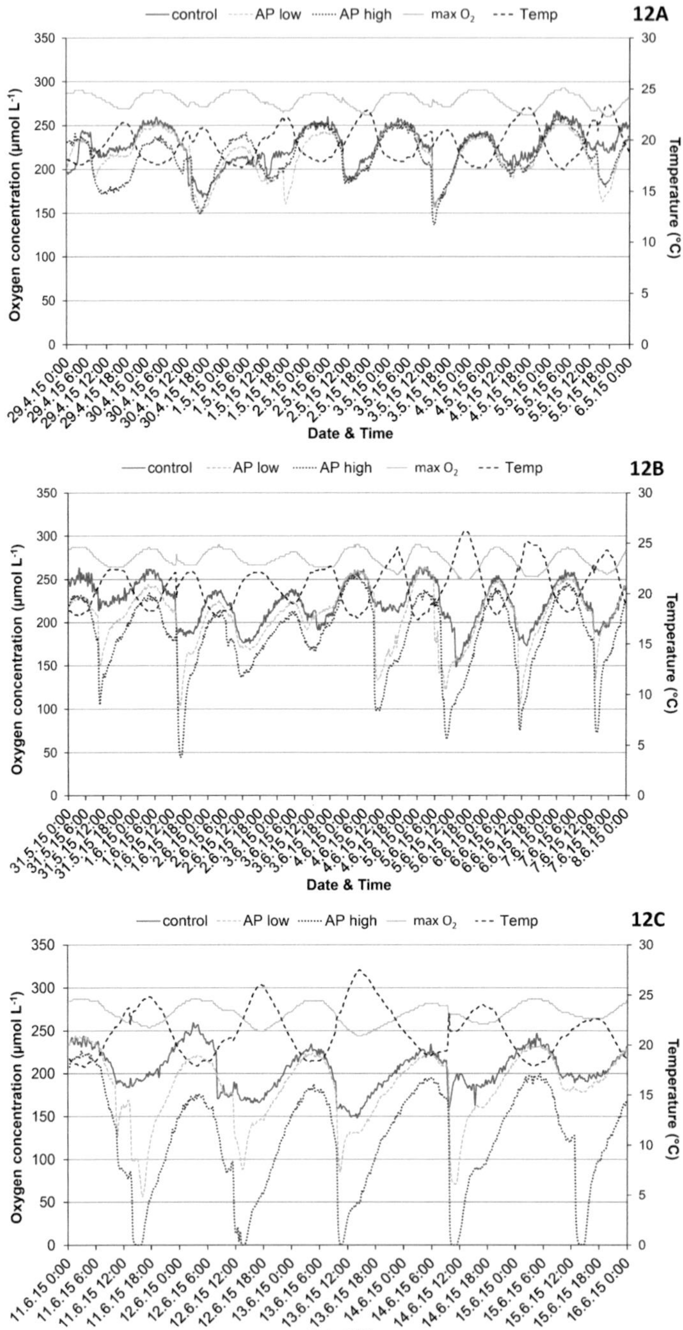
12A
control
AP low
AP high
max O2
Temp
Oxygen concentration (µmol L-1)
Temperature (°C)
Date & Time
12B
control
AP low
AP high
max O2
Temp
Oxygen concentration (µmol L-1)
Temperature (°C)
Date & Time
12C
control
AP low
AP high
max O2
Temp
Oxygen concentration (µmol L-1)
Temperature (°C)
Date & Time

Figure 11: Mean oxygen concentrations in the nutrient solution for the different treatments: control, prepared with well water (electrical conductivity (EC) 1.8 dS m^{-1}); aquaponics (AP) low (EC 1.8 dS m^{-1}) and AP high (EC 3.0 dS m^{-1}), prepared with fish waste water; and the theoretical maximal oxygen concentration at given temperature in pure water (max O_2). The temperature is the mean value of all sensors, installed in the different treatments. The oxygen concentration was measured every 15 min for three different measuring periods 12A, 12B, and 12C. The sensor measurement locations are marked in Figure 9. The plotted values are mean values of all sensors per treatment (control and AP high = each with 4 sensor, AP low = 12 sensor).

5.4.2 Correlations between oxygen concentration, photosynthetic photon flux density (PPFD), and temperature

Analysis of the correlation between the oxygen concentration in the nutrient solutions flowing in the cultivation trenches and PPFD showed that there is a negative correlation between both; the higher the radiation the lower the oxygen concentration. The correlation was highly significant ($p < 0.01$) in all considered measuring weeks (Table 9). The coefficient rose from the 1st measuring period ($r = -0.568$) to the 3rd measuring period ($r = -0.689$). A similar pattern was observed for the correlation between the oxygen concentration and temperature of the nutrient solutions. The relationship was also negative ($p < 0.01$) and increased over the measuring periods from End of April ($r = -0.607$; 1st period) until mid of June ($r = -0.817$; 3rd period). It was noticeable that the correlation between oxygen concentration and temperature was higher than between oxygen concentration and PPFD, although PPFD and the temperature were in turn significantly ($p < 0.01$) positively correlated. The positive correlation was very similar ($r = 0.562$ to 0.576) in all measuring periods.

Table 9: Pearson correlation coefficient between oxygen (O_2) concentration, photosynthetically active photon flux density (PPFD), and temperature.

	Measuring period			
Variable	**1st: 29.04. – 05.05.2015**	**2nd: 31.05. – 07.06.2015**	**3rd: 11.6. – 15.06.2015**	**All periods**
Mean temperature (°C)[1]	19.6	20.8	21.4	
Mean O_2 (µmol/L)[1]	232.1	215.4	186.3	
Mean PPFD (µmol m^{-2} s^{-1})	167.3	139.4	181.7	
O_2 * PPFD	-0.568**	-0.648**	-0.689**	-0.557**
O_2 * temperature	-0.607**	-0.784**	-0.817**	-0.773**
PPFD * temperature	0.562**	0.574**	0.576**	0.532**

The values for correlation were mean values of all treatments using all data points collected in the different periods.

[1]The data are mean values for each time point of all three treatments for day and night.

**The correlation is significant on a level of $p < 0.01$.

5.4.3 Oxygen consumption in the cultivation trenches

The oxygen consumption rate in the used nutrient solutions flowing in the cultivation trenches differed significantly between all treatments when 24 h values were considered (24 h; Table 10). When day and night were evaluated separately, the oxygen consumption rate at day and night was equal in control and AP low in measuring period A. The significant highest consumption during period A was found in AP high for all daytimes. In comparison to the control, the consumption was increased by 37.1% (24 h). Latter applied also for period B where the consumption was increased by 88.6% (24 h) in AP high in comparison to the control. Furthermore, in period B all treatments differ significantly in all evaluated times, and the oxygen consumption in the control was significantly lowest. The oxygen consumption in AP low was increased by 17.7% (24 h) compared to control.

The statistical analysis showed furthermore that the oxygen consumption was always significantly higher during the day than during the night (Table 10). The mean consumption during the day was 1.4 (AP high) to 1.7 (AP low) times higher during period A. In period B the differences between day and night increased and were 1.7 (AP high) to 2.8 (control) times higher during the day.

Table 10: Mean oxygen consumption rate in cultivation trenches, operate with nutrient solutions on basis of conventional hydroponics (control) and aquaponics (AP low and AP high) in two different measuring periods.

Measuring period	**Measuring time**[1]	**Oxygen consumption rate (µmol L^{-1} h^{-1})**		
		control	**AP low**	**AP high**
Period A (01.05.-03.05.2015)	24 h	79.8 ± 56.6^{b}	69.6 ± 36.5^{a}	109.4 ± 58.3^{c}
	Day	91.6 ± 59.6^{aB}	81.9 ± 36.3^{aB}	121.0 ± 55.5^{bB}
	Night	59.5 ± 44.4^{aA}	48.4 ± 25.7^{aA}	89.1 ± 57.6^{bA}
Period B (05.06.-09.06.2015)	24 h	58.1 ± 62.2^{a}	68.4 ± 82.8^{b}	109.6 ± 65.7^{c}
	Day	73.0 ± 60.9^{aB}	79.9 ± 90.9^{aB}	125.4 ± 64.6^{cB}
	Night	26.3 ± 52.4^{aA}	40.1 ± 49.4^{bA}	75.9 ± 54.6^{cA}

The data represents mean oxygen concentrations[8] during a period of three (period A) and five days (period B). Significant differences were analysed by analysis of Variance (ANOVA; normal distribution) and differences were tested using Dunnett-T3 test. Different superscript small letters indicate significant differences ($p < 0.05$) between the treatments within one measuring period. Significant differences between day and night within one treatment and per period were tested using Mann-Whitney *U* test and are indicated by different superscript capital letters. The control and aquaponics (AP) low were prepared with mineral fertilizer up to an electrical conductivity (EC) value of 1.8 dS m^{-1}, and AP high to an EC value of 3.0 dS m^{-1}. The nutrient solution for the control was prepared with well water, and the nutrient solutions for both AP treatments were prepared with fish waste water.
[1]Day = time between sunrise and sunset; night = time between sunset and sunrise; 24 h = mean value day and night.

5.4.4 *Oxygen consumption of the nutrient solutions and fish (waste) water*

In the experiments in which the decrease of oxygen levels in closed jars (jar experiment) with different solution samples (nutrient solutions, fish waste water, RAS water) were measured, it was observed that the water from the different treatments showed a different course of decreasing oxygen concentration over the evaluated measuring time (Figure 12). First of all it was noticed that three samples, fish waste water, AP high-in, and RAS water, showed a course deviating from that of the other samples. The oxygen concentration in the fish waste water dropped clearly the fastest. After a rapid initial reduction of the oxygen concentration in the first 21 min (Table 11), the decrease of the oxygen concentration became smaller. Only in the fish waste water treatment the oxygen concentration[9] dropped to 0 µmol L^{-1} within the measuring period of 370 min. Additionally, the nutrient solution from AP high-in showed as well a rapid decreasing oxygen concen-

[8] Taken from original publication; correct would be consumption
[9] Taken from original publication; correct would be consumption

tration after the measurement started. However, the reduction was less strong than in the fish waste water and levelled at around 160 µmol L^{-1} (Figure 12) after about the first 66 min (Table 11) of fast oxygen reduction. The oxygen concentration in RAS water dropped again less strong and levelled to a minimum of about 190 µmol L^{-1} at the end of the measuring period. All other treatments showed an equal behaviour. The oxygen concentration did not decline very strongly and levelled between 210 µmol L^{-1}and 240 µmol L^{-1} at the end of the measuring period.

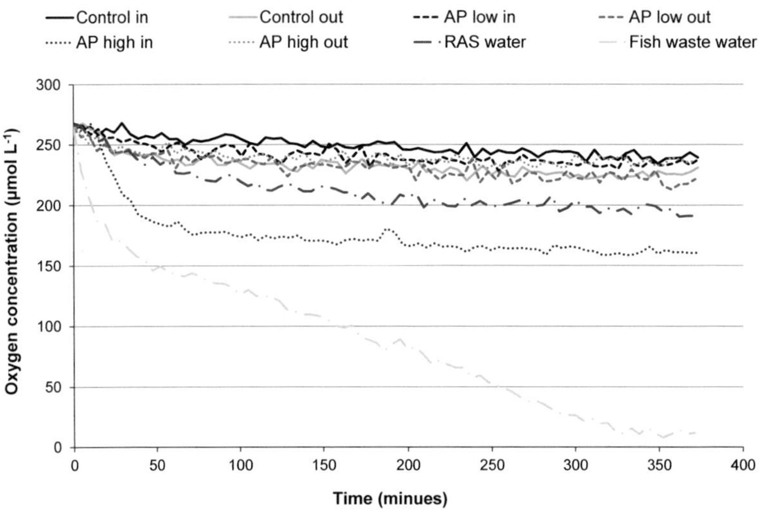

Figure 12: Oxygen concentrations in the jars filled with different nutrient solutions and fish (waste) water within a measuring period of 370 min. The nutrient solutions for the control were prepared with well water (electrical conductivity (EC) = 1.8 dS m^{-1}) and the aquaponics treatments (AP) were prepared with fish waste water (AP low = EC 1.8 dS m^{-1}; AP high = EC 3.0 dS m^{-1}). "In" and "out" behind the treatments describes the sampling point, described in materials and methods. Lines show average values for at least four jars.

The consumption rate ranged between 3.8 µmol L^{-1} h^{-1} (control in) and 36.2 µmol L^{-1} h^{-1} (fish waste water) when the total measuring period of 370 min was considered (Table 11). Considering the 370 min, the oxygen consumption in the fish waste water was significantly increased (36.2 µmol L^{-1} h^{-1}) in comparison to all other samples. The consumption was 3.1 times higher than in RAS water. The oxygen concentration of AP high-in and AP high-out differed by 11.8 µmol L^{-1} h^{-1}, and was significantly higher in AP high-in (17.0 µmol L^{-1} h^{-1}). For the other treatments (control and AP low) no difference in the oxygen consumption between the in and out going water in the trenches were detected.

As mentioned above, the nutrient solution prepared with fish waste water and mineral fertilizer for AP high and the pure fish waste water showed a high oxygen consumption rate in the first minutes (Table 11, Figure 12). The oxygen concentration of the inflow water of AP high decreased below the threshold oxygen concentration of 173 µmol L^{-1} within the first 66.4 min (period 1) and had a significantly higher oxygen consumption rate (83.5 µmol L^{-1} h^{-1}) than the mean value over the whole measuring period (17.0 µmol L^{-1} h^{-1}). After the threshold concentration of 173 µmol L^{-1} (period 1) was reached, the oxygen consumption rate dropped to 3.3 µmol L^{-1} h^{-1} for the rest of the time. The oxygen consumption rate in fish waste water reached 334.5 µmol L^{-1} h^{-1} in the first 21.4 min (period 1) and was also significantly higher when compared to the oxygen consumption rate during the whole measuring period (36.2 µmol L^{-1} h^{-1}) and the consumption rate after it reached the threshold concentration (24.4 µmol L^{-1} h^{-1}; period 2). Only in the latter both treatments (AP high-in and fish waste water) the oxygen consumption dropped below the threshold concentration of 173 µmol L^{-1} during the measuring period.

Table 11: Oxygen consumption rates of three nutrient solutions and fish (waste) water used to prepare the different aquaponic nutrient solutions, measured in 370 min (total), in period 1 and period 2.

Treatment	Total oxygen consumption rate (µmol L^{-1} h^{-1}) (370 min)	Oxygen consumption rate (µmol L^{-1} h^{-1}) in period 1	Minutes to reach period 2	Oxygen consumption rate (µmol L^{-1} h^{-1}) in period 2
Control in	$3.8 \pm 1.3^{a(3)A}$	-		-
AP low-in	$5.0 \pm 1.7^{ab(2)A}$	-		-
AP high-in	$17.0 \pm 1.3^{d(1)A}$	$83.5 \pm 10.5^{(1)B}$	66.4 ± 4.6	$3.3 \pm 1.6^{(1)A}$
Control out	$4.7 \pm 1.5^{ab(3)A}$	-		-
AP low-out	$6.6 \pm 0.8^{b(2)A}$	-		-
AP high-out	$5.2 \pm 1.7^{ab(2)A}$	-		-
RAS water	$11.5 \pm 1.6^{c(3)A}$	-		-
Fish waste water	$36.2 \pm 7.3^{e(3)AB}$	$334.5 \pm 99.9^{(3)C}$	21.4 ± 9.1	$24.4 \pm 5.2^{(3)A}$

The data represent the oxygen consumption rate during the whole measuring period (total) until the concentration reached the threshold concentration of 173 µmol L^{-1} (period 1), and after the concentration reached this threshold concentration (period 2). The mean values represent the mean of four (1), five (2) or six (3) repetitions ± standard deviation. Oxygen consumption was compared using Dunn-Bonferroni test and significant differences between the total oxygen consumption are indicated by different superscript letters ($p < 0.05$). Significant differences between all values (including until and after threshold concentrations) are indicated by capital letters. "In" and "out" behind the treatments describes the sampling point, described in materials and methods Whereas the nutrient solution for the control was prepared with well water (electrical conductivity (EC) 1.8 dS m^{-1}), the solutions for aquaponics (AP) were prepared with fish waste water (AP low = EC 1.8 dS m^{-1}; AP high = EC 1.8 dS m^{-1}).

5.5 Discussion

As presented in Figure 11, a continuous daily fluctuation of the oxygen concentration was detected in the cultivation trenches. The decreasing concentration during the day and increasing concentration during the night was found over all measuring periods from end of April until mid of June. However, it was noticeable that the intensity of the fluctuation increased clearly over time (Figure 11). Especially during the hot summer season in mid of June (Figure 11C) the oxygen concentration in the trenches dropped clearly over the day. The oxygen concentration in aquatic solutions in general, and in nutrient solutions in particular, is affected by different factors. It is generally known that the amount of soluble oxygen in aquatic solutions depends, among other things (pressure, salinity), on temperature. With increasing temperature the solubility of oxygen decreases (Carpenter, 1966).

This might be one factor of the day specific course and the decreasing oxygen concentration during the day. This was confirmed by the determined correlation between temperature and oxygen concentration ($r = -0.607$ to -0.817, $p < 0.01$) in the nutrient solutions (Table 9). Such high correlations were also found by Adams (1992) and Gislerød and Adams (1983). The temperature of the nutrient solutions in turn, of course, correlated positively with the radiation ($r = 0.562$ to 0.576, $p < 0.01$, Table 9). However, the correlation between the oxygen concentration and temperature was higher than between oxygen concentration and radiation. The typical daily course in the present study confirmed the observations by Holtman et al. (2009b). They found also a regular daily pattern with lowest oxygen concentrations during the afternoon when radiation was highest. However, the temperature dependence cannot explain totally the strong decrease of the oxygen concentration in the nutrient solution during the day, especially in AP high. The calculated maximal effect of the temperature on oxygen concentration was 33 µmol L^{-1} (29.04.-05.05.2015), 40 µmol L^{-1} (31.05.-07.06.2015), and 38 µmol L^{-1} (11.06.-15.06.2015). These values were clearly lower than the measured oxygen difference between day and night (Figure 11). In nutrient solutions some other factors have to be considered. As mentioned in the introduction, the plant itself affects the oxygen concentration due to the oxygen use for root respiration (Chun & Takakura, 1994; Gislerød & Kempton, 1983). For tomatoes, for instance, a root respiration rate of 3.24 µmol O_2 h^{-1} g^{-1} FM to 4.68 µmol O_2 h^{-1} g^{-1} FM was reported by Morard and Silvestre (1996). The root respiration itself is also temperature dependent (Jackson, 1980) and follows a time course with low rates during the night (low root temperature) and higher rates (high root temperature) during the day in consequence of temperature change (Palta & Nobel, 1989). For instance, a positive linear relationship between 9 °C and 20 °C was found for wheat by Morard and Silvestre (1996). In addition, microorganisms living in the root zone consume oxygen (Jackson, 1980; Strayer, 1994) and influence the oxygen concentration within the nutrient solution. Furthermore the growth and metabolisms of microorganisms in turn is temperature dependent and increases with increasing temperatures (Strayer, 1994). Both, the high activity of plants and microorganisms increased the oxygen consumption in the nutrient solution during the day significantly compared to the night (Table 10). The mean oxygen consumption during the day was 1.4 (AP high) to 1.7 (AP low) times higher in period A and 1.7 (AP high) to 2.8

(control) times higher in period B. Besides the higher plant and microorganism activity, the increasing root mass due to plant growth might also intensify oxygen depletion in the nutrient solution during the day and over the whole measuring period from end of April until mid of June (Figure 11). It was shown that a higher root weight results in an increasing oxygen depletion in nutrient solution (Morard & Silvestre, 1996).

During the measurement of the oxygen concentration in the cultivation trenches it was noticed that the oxygen concentration in the nutrient solutions of the different treatments differed increasingly over time (Figure 11). While the oxygen concentration was more or less similar in the first measuring period (29.04.-06.05.2015; Figure 11A), the depletion of the oxygen concentration during the day was stronger in aquaponics than in the control in the following weeks. Especially in AP high a continuous and strong oxygen depletion was measured over time. In the 3rd period (11.06.-15.06.2015, Figure 11C) the oxygen concentration even dropped to 0 in AP high, while the minimal oxygen concentration in AP low dropped to 57 µmol L^{-1} and in the control only to 148 µmol L^{-1} in the same period. The calculation of the oxygen consumption rate in the cultivation trenches confirmed these measurement results. However, it was detected that the oxygen consumption did not differ between control and AP low in period A (01.05.-03.05.2015), but did differ in period B (05.06.-09.06.2015) (Table 10). It was expected that the differences between AP low and control would have been increased further with passing time. In contrast, the oxygen consumption in AP high was always the significant highest in both periods. The higher oxygen consumption in aquaponics compared to control may be due to a higher microorganism activity and probably to higher amount of (soluble) solids. The decomposition of (soluble) solids consumes oxygen (Masser et al., 1992) and can contribute to oxygen consumption in the aquaponic nutrient solutions.

To investigate the effect of the nutrient solutions itself, without the influence of plants, on the oxygen consumption a further experiment was conducted (jar experiment; see paragraph 5.3.4). It was found that the oxygen consumption of the nutrient solution mixed with fresh water (control, "in" and "out") did not differ significantly to AP low-in, AP low-out, and AP high-out (Table 11). In the cultivation trenches the oxygen consumption of control and AP low was also not significant different in period A (day and night). However, in period B the differences were significant for the 24 h period and night, but did not differ during the day. The differences between control and AP low in terms of oxygen consumption within the cultivation trenches (Table 10, period B) could not be explained by the jar experiments (oxygen consumption of nutrient solutions itself). Thus, the differences in oxygen consumption in the trenches between control and AP low were probably not caused by a difference in oxygen consumption of the nutrient solution itself, but more likely by plant related factors (root respiration and/or respiration of microbes attached to the root surface). However, ANOVA revealed a significant effect on the oxygen consumption in AP high-in. However, in general it was interesting that the oxygen depletion in AP low and control was relatively similar, even if AP low was prepared with the same fish waste water as AP high. Therefore, a more similar oxygen consumption, whether in the cultivation trench or in the jars experiment, for both aquaponic treatments were expected. Nevertheless, the significant increased oxygen consumption in AP high-in indicated that the stronger depletion of oxygen, measured in the cultivation trenches (see above) was probably mainly due to a high oxygen consumption of the nutrient solution itself. This in turn could be caused by using fish waste water, what for its part showed the highest oxygen depletion and consumption (Figure 12, Table 11). However, the latter one would also have increased the oxygen consumption of AP low. Both nutrient solutions, AP low and AP high, differed just in the amount of added fertilizer salt. According to Bloom and Epstein (1984) a moderate increased salt content can increase root respiration, especially in salt sensitive cultivars. As such, the higher salt concentration in AP high might be at least partly responsible for higher oxygen consumption detected in cultivation trenches. For the experiment all samples were saturated at the beginning and the start (saturation) concentration within the samples did not differ (data not shown) even if the salt concentration does have an effect on oxygen solubility (MacArthur, 1916). In addition, it was noticed that only in the AP high nutrient solution, the oxygen consumption dropped significantly by 69.4% from the inflow (17.0 µmol L^{-1} h^{-1}) to the outflow (5.2 µmol L^{-1} h^{-1}). However, in this context it must be considered that only one measuring (point in the cultivation trenches) with two sensors was installed at the inlet and the outlet, respectively, in AP high and control. In AP low, three measuring points with six sensors were installed at the inlet and outlet. The different number of sensors might influence the calculation and results should be interpreted with care. Likely, something changed within the nutrient solution of AP high as it passed the root zone. The reduction of solids by the root mass may play a role, but this would also be the case for affecting the oxygen consumption of AP low-out compared to AP low-in. However, the latter was not determined. Further experiments are necessary to investigate the differences between AP high and AP low.

The differences between RAS water and fish waste water are of further importance. The results showed that something happened to the fish water during its transfer from the RAS to the storage tank. While the oxygen consumption of the RAS water was 11.5 µmol L^{-1} h^{-1} during the measuring period of 370 min, it was 3.1 times higher in the fish waste water. Furthermore, oxygen consumption[10] within the RAS water dropped never below 173 µmol L^{-1}. In contrast, the oxygen concentration in the fish waste water dropped very fast within the first 21.4 min below 173 µmol L^{-1} (consumption rate = 334.5 µmol L^{-1} h^{-1}, period 1) and afterwards slower (consumption rate = 24.4 µmol L^{-1} h^{-1}, period 2), but as compared to the other samples still fast, to nearly zero. As shown in Figure 8 and described in materials and methods, the water passed a 3-chamber-pit (3-cp) before it was used to prepare the nutrient solution. As described by Suhl et al. (2018c) it was detected that in the 1st chamber probably denitrification processes occurred due to sludge accumulation and a low oxygen environment. However, a microbiological change may occur in the water as well. It seems reasonable that the number/density of denitrifying bacteria increased in the water passing the 3-cp. Many denitrifying bacteria, e.g. *Pseudomonas* ssp. *and Paracoccus* ssp. which are also present in RAS (Leonard et al., 2000; Sugita et al., 2005), are facultatively anaerobic and can survive and grow also under aerobic conditions in the presence of oxygen (Davies et al., 1989; Patureau et al., 2000). Even if the denitrification is reduced under aerobic conditions, it does not stop totally (Patureau et al., 1996) and depends on growth conditions (Davies et al., 1989). A higher number of bacteria might explain the higher oxygen consumption in the fish waste water, when compared to the RAS water.

Many studies were carried out to investigate the effect of oxygen deficit on plant growth. But well-defined threshold concentrations for oxygen concentrations in nutrient solutions are rare. Schröder and Lieth (2002) recommended in general a level above 60% saturation. Below 60% saturation the growth of barley seedling roots was inhibited (Jackson, 1980). However,

[10] Taken from original publication; correct would be concentration

Goto et al. (1996) found no negative effect on lettuce growth in hydroponics from a concentration of 2.1 mg L^{-1} (65.6 µmol L^{-1}) and higher (25% of saturation at 24 °C). Jackson (1980) found that the negative effect on leaves occurs at lower concentrations than to roots. While the roots were inhibited below 60% saturation, the leaves were affected only below 10% saturation. According to Gislerød and Kempton (1983) the critical oxygen concentration for cucumber growth is 3 ml L^{-1} (93.8 µmol L^{-1}). Zeroni et al. (1983) stated a critical oxygen value of 50% of oxygen saturation (≙ 125 µmol L^{-1} to 150 µmol L^{-1}) for tomato fruit production, but their recommendations for tomato growth and fruit production is an oxygen concentration of 65% (≙ 162.5 µmol L^{-1} to 193.8 µmol L^{-1}) of oxygen saturation (Zeroni et al., 1983). However, considering of the lower critical oxygen concentration of 4.0 mg L^{-1} (≙ 125 µmol L^{-1}) for tomatoes, the oxygen concentration in the cultivation trenches fell part clearly below the critical concentration in AP high and AP low (2nd and 3rd period; Figure 11b, and c). Despite the depletion of the oxygen concentration to 0 µmol L^{-1} in AP high during high summer season in the present study (Figure 11C), the fruit yield and plant growth was not clearly affected (Suhl et al., 2018a). The vegetative growth (number of leaves, total plant length) was similar in all three treatments. The same applies to the generative development (fruit settings, total number of trusses). However, the fruit yield was significantly higher in AP high compared to AP low but not in comparison to the control. The higher yield in AP high was due to a heavier fruit weight (Suhl et al., 2018a). After the strong oxygen depletion detected in the nutrient solution in summer time an air bubble stone was installed in each nutrient solution tank to prevent successfully (data not shown) further strong oxygen depletion during daytime. That the lack of oxygen did not affect the plant growth obviously might be on the one hand due to the short measuring period of absolute lack of oxygen. According to Morard and Silvestre (1996) short and temporary oxygen deficits do not induce irreversible nutritional stress in plants. It was also reported that the ornamental plants *Ficus* and *Chrysamthemum* can probably adapt to constant low oxygen concentrations. Soffer et al. (1991) found that the differences in vegetative growth of both species between low and high oxygen concentrations in the nutrient solution were higher after 47 days than after 88 days. They explained this by presence of an adaptation mechanism. It might be that also the tomato plants in the present study adapt to regularly happening oxygen deficits. On the other hand, in NFT the root mass is usually not submerged completely in the nutrient solution. A part of the roots are above the water surface and can absorb oxygen from ambient air. A further indication that the lack of oxygen in the present study did not affect the plant growth was that the mineral content in the fruit was partly similar (Mg) to the control or even higher (Zn, K) in AP high (Suhl et al., 2018a). It was reported that oxygen depletion results in a decreased nutrient uptake (Morard et al., 2000; Morard & Silvestre, 1996; Stępniewski & Przywara, 1992).

The use of DRAPS is promising for saving water and nutrients but also needs new insights in the crop management requirements in the system. Our results clearly demonstrate that the use of the fish water in tomato cultivation has an impact on the oxygen level in the root zone. Depending on the time of day, as well as the crop developmental status and season the oxygen levels in the cultivation trenches reach very low values when fish waste water was used. These low levels may affect the yield and quality of the crop, as well as make the crop more vulnerable for pathogens (Chérif et al., 1997). In the present study no negative effects on plant growth were detected due to mentioned reasons. Nevertheless, should oxygen depletion occur in NFT, intermitted nutrient solution flow might be an option to guarantee adequate oxygen supply to the roots. Intermitted means an alternation of irrigation and dry period (Economakis, 1993). However, using fish waste water in other hydroponic systems can be problematic. In deep water culture systems, mainly used for leafy vegetable production, the roots are completely submerged in nutrient solution. In these systems the adequate oxygen is in general a challenge and aeration is necessary (Jensen, 1999; Zeroni et al., 1983). In those systems, the use of fish waste water could intensify oxygen depletion. According to the present results it is recommended, especially for aquaponics, to control the oxygen concentration within the hydroponic system to react in a timely manner when oxygen depletion occurs. However, a modified transfer of the fish water directly from the RAS to the storage tank without passing the 3-cp as described by Suhl et al. (2018b), would probably prevent strong oxygen depletion in the nutrient solution prepared with fish waste water. Another possibility to cope with oxygen depletion might be to introduce an UV irradiation device for fish waste water for disinfection to lower drastically the amount if microorganisms.

5.6 Conclusion

It was demonstrated that fish waste water can have a high oxygen consumption rate which can result in low oxygen concentrations in nutrient solutions in NFT systems. Using fish waste water as in double recirculating aquaponic systems (DRAPS) increased the risk that the oxygen concentration fell to zero during daytime in summer. This is probably due to a higher microorganism activity and a higher content of (soluble) solids within fish waste water as compared to fresh water. However, also the transfer and storage of the fish waste water affects the oxygen consumption of the pure nutrient solution. After transfer and storage the oxygen consumption increased considerably. Additionally, the rate of fertilization of the fish waste water seems to have an effect on the oxygen consumption of the nutrient solution. It seems that a higher EC in the nutrient solution increased the oxygen consumption of the pure nutrient solution. In summary, when fish waste water is used to irrigate hydroponically grown plants a control of the oxygen concentration is highly recommended, especially at high electrical conductivity (≥ 3).

6

An innovative suction filter device enhances double recirculating aquaponic systems by preventing nitrogen loss

Johanna Suhl, Dennis Dannehl, Daniela Baganz, Uwe Schmidt, and Werner Kloas

Aquaculture Engineering *82, (2018) 63–72*
DOI: 10.1016/j.aquaeng.2018.06.008

6 An innovative suction filter device reduces nitrogen loss in double recirculating aquaponic systems

6.1 Abstract

The increase of global population causes rising food demand and concomitantly scarcity of resources. These facts including global warming need the further development of resource efficient food production systems such as double recirculating aquaponic systems (DRAPS). This recent technology is suitable for intensive food production with minimal resource input and maximal reduction of emissions. In DRAPS fish water can be adjusted for plant growth (pH, nutrients) and both species are produced separately under optimum conditions.

Primarily, the present study was conducted to investigate the improvement of DRAPS by implementation of an innovative suction filter device and, in addition, it is the first empiric investigation of combined African catfish (*Clarias gariepinus*) and tomato (*Solanum lycopersicum* L., cv. Pureza) production in DRAPS.

Due to an immense loss of nitrogen caused by the primary sedimentation unit as connection between the fish and plant parts, the original DRAPS was modified by a new developed suction filter device to remove solid particles at minimal nitrogen loss. The nitrogen loss was significantly reduced by that system modification (≈43%) and thus resulted in a quite large reduction of greenhouse gas emissions due to the reduced need for manufacturing of nitrogen fertiliser. The combined production of African catfish and tomatoes in aquaponics revealed similar fruit yields of tomatoes as for the control using artificial fertiliser, while the overall fertiliser use in aquaponics was about 13.2–77.7% lower. However, fertiliser saving was highest after modification and replacement of the sedimentation unit by the new suction device. In addition, in comparison to already existing aquaponic systems, the total biomass output was significantly increased. The results were evaluated under consideration of intensive crop production in hydroponics.

6.2 Introduction

Climate change, water and fossil fuel scarcity as well as food shortage is linked to an increasing demand of water, food and energy due to growing world´s population (Godfray et al., 2010). In a life cycle assessment study for tomato production it was demonstrated that fertilisers cause by far the greatest environmental impact (Torrellas et al., 2012). Especially during the manufacturing of nitrogen fertiliser, high emissions of CO_2, N_2O, NO, and NO_2 occur (Wood & Cowie, 2004), which have major impacts on global warming, eutrophication, and air acidification (Antón et al., 2004; Torrellas et al., 2012). Therefore any attempt to make use of nitrogen within value added chains of fertilisers supports sustainability by reducing environmental impacts. Furthermore, agriculture uses 70% of the global fresh water resources (FAO, 2011) and thus freshwater availability is one of the most important issues for maintaining food security. As such, the fresh water consumption by, for instance, tomato production is an issue as well. Depending on the season and production systems the water footprint of tomato production in greenhouses can rise to 122.6 L kg^{-1} produced tomatoes (Almeida et al., 2014). In order to achieve a high food quantity without adverse impacts on the environment, innovative and sustainable food production systems are urgently needed (Godfray et al., 2010).

Aquaponics, which is the combined production of aquaculture and plants in hydroponics, is a sustainable option to reduce fertiliser and water use (Maucieri et al., 2017). The main advantage of aquaponics is the double use of resources like water and nutrients, used firstly for fish and secondly for plant production (Rakocy et al., 2006; Tyson et al., 2011). Due to metabolic processes, fish excrete mainly ammonia (NH_3) which is usually protonated immediately to ammonium (NH_4^+) after it enters the water. A good overview of the biological background is given by Wongkiew et al. (2017). While NH_3 is highly fish toxic, nitrate (NO_3^-) is much less harmful to fish (Timmons et al., 2010). For intensive fish production in recirculating aquaculture systems (RAS) NH_4^+ is converted into NO_3^- in a biofilter by nitrifying bacteria. Nitrate nitrogen (NO_3-N) concentrations in fish cultures can be up to 500 mg L^{-1} (Honda et al., 1993) and are commonly controlled by daily water exchange (Timmons et al., 2010). The release of such nitrogen rich fish waste water into the environment can result in a massive environmental pollution (d'Orbcastel et al., 2009). In order to reduce the overall emissions of aquaculture and fertiliser manufacturing it is absolutely appropriate and sustainable to use the nutrients from fish water for plant nutrition in aquaponic systems, thus creating value added chains.

Conventional single recirculating aquaponic systems (SRAPS) are mainly operating as a one loop low-tech system. Due to the different environmental requirements of fishes and of plants, compromises in terms of water quality have to be made and reveal reduced yields compared to separate single production systems for fishes and plants, respectively (Graber & Junge, 2009; Vergote & Vermeulen, 2012; Wortman, 2015). Therefore, an innovative approach to overcome these problems was recently published by Kloas et al. (2015). They developed a high-tech double recirculating aquaponic system (DRAPS), where two independently operating systems, a RAS and a recirculating hydroponic system, are coupled. The connection of both is realised by a one-way-valve creating a unidirectional water flow (Kloas et al., 2015). This means that the main specific characteristics of DRAPS are the separation, the independent operation of both productions cycles (fish and plant), and the unidirectional water transfer from fish to plants. Latter means that the water is not transferred back from the hydroponic system to the RAS as known from SRAPS. Thus, the growth conditions can be optimised for each species separately and allows an intensive production of fish and plants comparable to single production units (Kloas et al., 2015; Suhl et al., 2016). Furthermore an indirect water transfer from the hydroponic unit back to RAS can be implemented for the full version of DRAPS (Kloas et al., 2015) by regaining the evapotranspirated water in the greenhouse by an active cooling unit in order to close the water loop. A good overview about the advantages and disadvantages of DRAPS compared to SRAPS is given by Suhl et al. (2016). Since both, the RAS and the hydroponic system, are working independently from each other in DRAPS, the water quality can be optimised for each individual cycle in order to produce fish and, for example, tomatoes in an intensive way. So far, only tilapia, a common aquaculture fish, combined with tomato plants have been produced using DRAPS (Kloas et al., 2015; Monsees et al., 2017b; Suhl et al., 2016). Delaide et al. (2016) also used the idea of DRAPS to produce lettuce with waste water from a tilapia rearing system. According to the fish species chosen and the respective rearing conditions, the waste water quality of RAS can vary widely in terms of nutrient concentrations and composition. In order to evaluate the use of DRAPS for further fish species, the present study focused on the first combined production of the African catfish and tomatoes. However, the main focus of the recent study was the replacement of the yet used sedimentation unit by an innovative filter suction device in the RAS part of the DRAPS to improve its efficiency and sustainability. Thus in the already established DRAPS (Suhl et al., 2016) it has been investigated whether the innovative filter suction device can provide fish waste water with a higher nitrogen concentration for intensive tomato production and thus reduce the environmental impacts. Yield quantities in terms of fish and tomatoes, the fertiliser use efficiency and the fresh water use per unit biomass of fish and tomatoes were determined for a system comparison considering DRAPS and a conventional separate hydroponic unit serving as control. In this context the reduction of emissions caused by saving nitrogen fertiliser due to the use of the modified DRAPS compared to the original DRAPS and the control was estimated. To evaluate the prospects of fish and fruit productivities, the optimum plant to fish ratio and its potential total output were calculated.

6.3 Materials and methods

6.3.1 Experimental set-up

The DRAPS used in the present study was described in detail by Suhl et al. (2016). A spatially separated RAS contained four identical glass fiber tanks with a total net production volume of 7.2 m^3, a mechanical sedimentation filter (1.3 m^3) to clean the fish water from solids, a trickling biofilter with pump sump (2.3 m^3) and a reception tank. The latter collected the water after passing the biofilter, before it was pumped back to the fish tanks. The RAS had a total water volume of about 12.0 m^3. To keep the water quality on an appropriate level for fish rearing, a regular water exchange was conducted. The daily water exchange rate over the whole production period was in average 8.5% of the net fish production volume. This water exchange occurred until calendar week (CW) 29 at the mechanical sedimentation filter (Figure 13A). The waste water was drained there to clean the filter up from sludge. During this procedure, the released waste water was discharged into a 3-chamber pit and the drained water was replaced by fresh water. The mechanical sedimentation filter plus the 3-chamber pit is referred to as sedimentation unit (Figure 13A). The 1st chamber of the 3-chamber pit was used for sludge sedimentation. From chamber 1 the water flowed passively by overflow into the 2nd and following into the 3rd chamber. The aeration in chamber 2 was realised using a fan turned on 24 h a day. From chamber 3 the fish waste water was pumped into a storage tank (1 m^3) placed within the greenhouse. It was kept there until its use as process water for mixing up the nutrient solution for hydroponic plant production. This described system is hereinafter referred to as "original DRAPS" (Figure 13A).

In CW 29 the DRAPS was modified. The reason for this modification was based on the inefficient performance of the 3-chamber pit resulting in nitrogen losses, which decreased drastically the nitrogen availability for plant production. As such, a new suction filter device for purifying the recirculating fish water (PAL KLAR; Patent number DE 102014004767A1) was implemented into the RAS and replaced the sedimentation unit (Figure 13A and B). The functionality of the suction device is described below in section 6.3.2. For the water exchange, the fish waste water was pumped several times a day to the storage tank in the greenhouse without passing the 3-chamber pit and was replaced by fresh water. The system with the implemented new filter system is hereinafter referred to as "modified DRAPS".

The described RAS was connected to a Venlo-type greenhouse with South-North orientation and a gross area of 139 m^2 as described by Suhl et al. (2016). The experimental layout is described in paragraph 6.3.2.2.

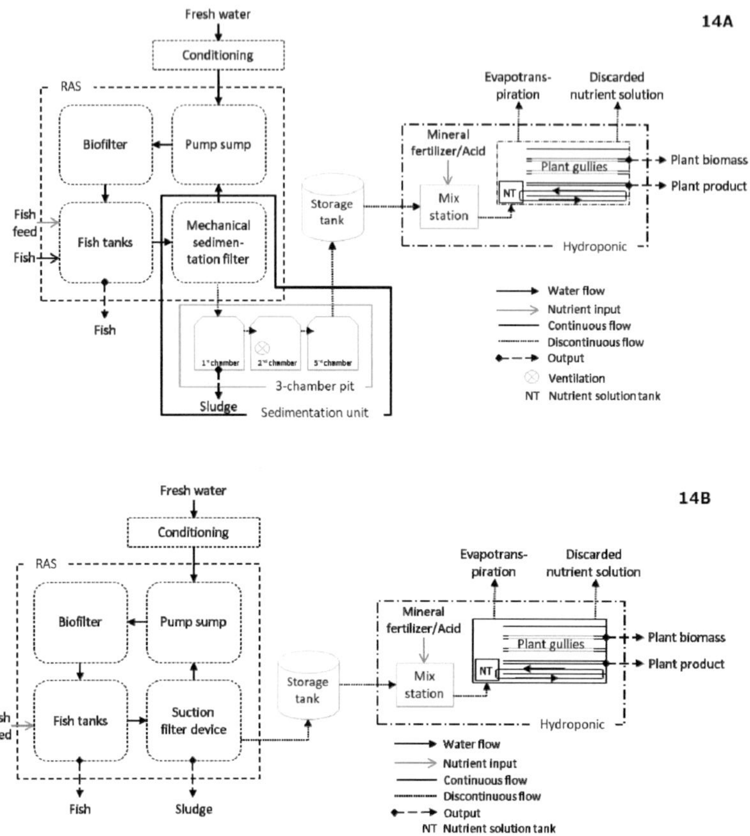

Figure 13: Schematic flow diagram of the original (14A) and modified (14B) double recirculating aquaponic system (DRAPS) with main in- and outputs. The blue lines mark the recirculating aquaculture system (RAS) with fish-rearing tanks, mechanical sedimentation filter, pump sump and biofilter; the green tags comprise the hydroponic unit with the nutrient solution tank (NT) and the plant gullies. The RAS and hydroponic unit are connected via a 3-chamber pit (original DRAPS), a storage tank and the mix station to adjust the fish waste water using mineral fertiliser and acid. The mechanical sedimentation filter and 3-chamber pit in the original DRAPS (14A) form the sedimentation unit.

6.3.1.1 The suction filter device

The first prototype of the new suction filter device consists of a combination of a stainless steel sieve and a sucker to remove the solids continuously from the fish water. In detail, the filter system was implemented within a separate water tank (700 L) between the fish rearing tanks (inflow) and the pump sump (outflow; Figure 13B). Between in- and outflow, the stainless steel sieve (mesh width 150 µm) moved in a circular motion, while one half of the sieve was above and the other half of it was below the water surface. While the fish water passed the sieve the sludge stuck on it. Above the water surface a suction tube was installed in front of the sieve. The connected wattage of the suction device was 4.5 kW. To remove the solids from the mesh, the tube moved continuously on a linear glide rail from the outer edge to the middle of the sieve. The speed of the tube movements was controlled by water level.

6.3.2 Fish and tomato production

The experiments were conducted from 27th January 2016 until 31st December 2016 (CW 4–52). During this period African catfish (*Clarias gariepinus*) and tomato (*Solanum lycopersicum* L., cv. Pureza) were produced in combination. While the tomato production was stopped already on the 20th October 2016 (CW 42), the fish production was still running until the end of the year (CW 52).

6.3.2.1 Fish production

The RAS was stocked stepwise until CW 16 when all four fish tanks were stocked with fish. From CW 17 until the end of the year only three of four fish tanks were coupled to the system. One fish tank was always in retention period (see below). The mean stocking fish weight was 332.8 g fish^{-1} and started with an initial mean stocking density of 28.6 kg m^{-3}. The fish were fatted for around 86 days up to a mean slaughter weight of 1.4 kg. Before the fish were harvested, the feeding was stopped and the respective fish tank was decoupled from the RAS (retention period; 8–35 days).

The fish production was separated into three phases (Table 12): (1) the start-up phase, where fish production started slowly and the stocking steadily increased (CW 4–20), (2) the 2nd period (CW 21–29) started with reaching the steady state of fish rearing and the original DRAPS was used, and (3) the 3rd period (CW 30–52), where the modified DRAPS with the new suction filter device was used (see 6.3.1). The African catfish were fed with ME-4.5 Meerval 44-14 (Skretting, France) containing 44.0% crude protein, 14.0% crude fat, 22.3% carbohydrates, 10.5% crude ash, 1.2% crude fiber and 18.9 MJ kg^{-1} digestible energy. The daily mean feeding rate was 1.3% of fish weight and the feed was given four times a day.

To evaluate the fish production, the total weight gain (final stocking weight minus initial stocking weight) per fish charge was calculated. Additionally, the feed input was measured and used to calculate the feed conversion ratio (FCR, kg feed input divided by total weight gain), expressed as kg kg^{-1} for each fish charge.

Table 12: Fish production and water sampling phases.

Phase	Description	Calendar week	
		Fish production	**Water sampling for analysis**
1	Start-up phase (increasing fish growth and stocking)	4–20	8–20
2	Original DRAPS (sedimentation unit) (Figure 1A)	21–29	21–29
3	Modified DRAPS (new suction device) (Figure 1B)	30–52	35–39

6.3.2.2 Plant growth conditions

Two single and three double gullies were placed in the experimental greenhouse where only the double gullies, arranged for three repetitions per treatment, were used for experiments. The nutrient solution for the plants was kept in a 150 L tank at the end of each gully. As such two different nutrient solutions were mixed using different process waters. The nutrient solution for the control was mixed with fresh water, and the one for the aquaponics with fish waste water from RAS as described in section 6.3.1. Both nutrient solutions were adjusted with mineral fertiliser to an EC (electrical conductivity) level of 2.6 dS m^{-1}. The target nutrient solution composition for both was identically as follows: pH=5.6 (adjusted with phosphoric acid); 241 mg L^{-1} nitrogen (N), 46.5 mg L^{-1} phosphorus (P), 371.5 mg L^{-1} potassium (K), 216.5 mg L^{-1} calcium (Ca), 58.3 mg L^{-1} magnesium (Mg), 47.2 mg L^{-1} sulphur (S), 0.8 mg L^{-1} iron (Fe), 0.3 mg L^{-1} boron (B) and zinc (Zn), 0.05 mg L^{-1} copper (Cu) and molybdenum (Mo) and 0.6 mg L^{-1} manganese (Mn) according to De Kreij et al. (1999). The nutrient solution was mixed by an automatic mixing station controlled by EC and pH, which mixed the process water with the nutrient stock solutions and acid (Figure 14). The stock solutions were prepared according to the current nutrient concentration of the respective process water and the target nutrient solution for plant production. For preparing the nutrient stock solution the standard fertilisers calcium ammonium nitrate (CAN; containing 1.1% ammonium), potassium nitrate (KNO_3), magnesium sulphate ($MgSO_4$), potassium hydrogen phosphate (KH_2PO_4), and ready to use trace elements solution (YaraTera Tenso Cocktail, Yara UK Limited; N E Lincolnshire, United Kingdom) were used. To provide optimal nutrient concentrations, the nutrient concentrations of the process water were analysed periodically (see 6.3.5). The first use of fish waste water from the RAS to mix the nutrient solution occurred in CW 11.

Each gully had a total of 24 tomato plants where the plants were planted when the first inflorescence became visible. Since the treatments were applied in three randomised repetitions, a total of 72 plants per treatment were used. Based on a total net-acreage of 83.5 m^2 and under consideration of all plants, a plant density of 2.3 plants m^{-2} was achieved. The nutrient supply was ensured by a recirculating system applied as nutrient film technique. The nutrient solutions were pumped with a rate of 9 L min^{-1} for 24 h per day into the respective gullies. To ensure an adequate oxygen concentration in the nutrient solution an air stone, blowing air into the nutrient solution, was installed in each nutrient solution tank.

To maintain the desired climatic conditions for plant growth, ventilation was opened above 21 °C to cool the greenhouse and the setpoints of the heating system were defined as 19 °C and 17 °C for day and night, respectively. The installed energy screen was closed below 3 W m^{-2} and was opened stepwise after sunrise to warm up the cold air in the roof region. Furthermore, high pressure sodium lamps were switched on at a global radiation below 20 W m^{-2} between 7 a.m. and 8 p.m. until tomatoes reached maximal hanging height.

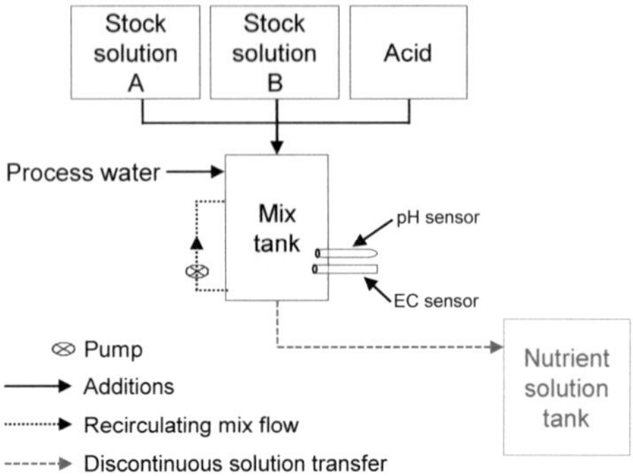

Figure 14: Schematic flow diagram of the automatic mixing station for preparing nutrient solution with main in- and outputs. The mixing process was controlled by electrical conductivity (EC, 2.6) and pH (5.8).

6.3.3 Determination of tomato yield

The tomato fruit were harvested once a week from CW 15 until CW 42. To compare the yield of both treatments, the harvested tomatoes of each row were weighted. For evaluation, only healthy fruit were considered and hereinafter referred to as marketable fruit yield. The results were expressed as kg cumulative yield per plant at the end of the experiments.

6.3.4 Sampling, preparation and chemical analysis of nitrogen in fish water (RAS water) and fish waste water

Between CW 8–14 every third week, and in the following periods every second week, samples of fish water and fish waste water were taken from the reception tank in the RAS (hereinafter referred to as RAS water) and the storage tank in the greenhouse (hereinafter referred to as fish waste water), respectively, in order to investigate the nitrogen levels. Until nitrogen analysis the samples were stored at −20 °C. As such, the ammonium nitrogen (NH_4-N) and nitrate nitrogen (NO_3-N) concentrations in the water samples were analysed using continuous flow analysis (San++, Fa. SKALAR; Breda, Netherlands). These measuring methods are exactly described by Suhl et al. (2016). The N_{total} was calculated as sum of NH_4-N and NO_3-N. The water samples were analysed in duplicate and the results were expressed as mg L^{-1}. To evaluate the results, data were separated into three phases that differed slightly from the periods described in 6.3.2.1 (Table 12). The 1st phase started only at CW 8 with the first sampling and is defined as period of CW 8–20. The 2nd phase is described similarly as CW 21–29 and the 3rd phase represents the measuring period between CW 35–39. To examine the N loss due to the 3-chamber pit, in the 2nd phase water samples were taken from the 2nd and the 3rd chamber of the 3-chamber pit using the original DRAPS.

6.3.5 *Calculation of total water and fertilizer use and its efficiencies*

6.3.5.1 Water and fertilizer use

The total fresh water volume added to the RAS was measured by a water metre. The water volume of fresh water as well as fish waste water used to mix the nutrient solutions was measured by a fillinglevel metre which was implemented in the automatic mixing tank. The data were cumulated during the experiment to evaluate water and nutrient saving. The total volume of added fresh or fish waste water, respectively, was calculated in L plant^{-1}. The mixing station also measured the amount of added nutrient stock solution to the different plant treatments. Based on this data, the amount of total mineral fertiliser used per L nutrient solution was calculated and expressed as g L^{-1}. To compare the data, it was calculated as mean value per month from March onwards. This starting point was chosen because the fish waste water was firstly used for mixing the nutrient solution since mid of March (see 6.3.2.2). The loss of nutrients caused by regularly discarded volumes of nutrient solution was not considered in these calculations.

6.3.5.2 Fresh water and fertilizer use efficiency

The fresh water use efficiency for the control was calculated as the ratio of marketable fruit yield and total fresh water used to mix the supplied nutrient solution. The results were expressed as kg tomato fruit per m^3 water supply (kg m^{-3}). For the aquaponic treatment, the total fresh water use for fish production in the RAS were considered to calculate how many fish and tomatoes were produced per m^3 fresh water input.

For the calculation of the fertiliser use efficiency the total marketable fruit yield was divided by the sum of supplied mineral fertiliser (in kg per plant) to the plant treatments. The results were expressed as kg fruit yield per kg fertiliser supply (kg kg^{-1}).

6.3.6 *Statistical analysis*

To evaluate the effects of nutrient solutions based on fresh water and fish waste water on all mentioned parameters the SPSS package version 19.0 was used. The yield, the fertiliser addition, the water consumption, as well as the fertiliser use efficiency and the fertiliser supply per month determined for the control and the aquaponic treatment, respectively, were compared using parametric t-test after normal distribution has been confirmed by Shapiro-Wilk test. If normal distribution failed the non-parametric Mann-Whitney *U* test was used.

In terms of normal distribution the nitrogen concentrations in the different waters was compared using univariate analysis of variance and in the following tested using Tukey-HSD (homogeneous variances) or Tukey-Kramer (unequal sample size) and Dunnett-T3 (heterogeneous variances) test. Non normal distributed data were compared using Kruskal-Wallis test followed by Dunn-Bonferroni test. The calculated mean values were displayed in figures as bars or in tables as numbers ± standard deviation and significant differences ($p<0.05$) between the treatments were indicated by different small superscript letters.

6.4 Results and discussion

6.4.1 *Fish and tomato fruit production*

It was already stated by Kloas et al. (2015) and Suhl et al. (2016) that double recirculating aquaponic systems (DRAPS) can compete with separated tilapia or tomato productions. In comparison to conventional low-tech single recirculating aquaponic systems (SRAPS), DRAPS are developed for intensive large-scale and sustainable food production. As such, high productivities in both units, fish and plants, are required.

The evaluation of fish productivity in the used DRAPS, producing African catfish and tomato, showed, that the mean stocking density in the RAS (excluding the fish tank in retention period) rose during the start-up period from 27.3 to 128.9 kg m^{-3}. During the remaining time of fish rearing, the mean stocking densities varied between 42.6 and 124.5 kg m^{-3} while the final stocking density at harvesting time varied between 85.0 and 185.5 kg m^{-3} (Table 13). The final stocking density was not as high as reported by van de Nieuwegiessen et al. (2008) for conventional African catfish production in RAS. Considering the entire production period, the mean final stocking density was 112.9 kg m^{-3} in the present study (Table 13), whereas the final stocking densities ranges between 200 to 500 kg m^{-3} in conventional RAS production in the Netherlands (van de Nieuwegiessen et al., 2008). The comparatively low stocking density was mainly chosen to keep the water exchange rate at a relative low level because the water demand of the tomatoes was restricted by the net acreage. The total weight gain of African catfish during the whole production period (CW 4–52) was 1730.8 kg (Table 13), which was achieved by feeding 1557.3 kg fish feed. Thus, a mean feed conversion ratio (FCR) of 0.91 was realised. The FCR never exceeded 1.0 and varied between 0.81 and 0.96. The calculated mean FCR of 0.91 reflected an appropriate growth performance of the African catfish and was equal to that reported by van de Nieuwegiessen et al. (2009). In this context, published data about African catfish production in aquaponics is rather limited. However, it was demonstrated that the FCR in the used DRAPS was lower than reported by Endut et al. (2016) indicating a FCR between 1.13 and 1.29 for African catfish produced in SRAPS. The higher FCR in terms of the mentioned studies might be caused by suboptimal rearing conditions and a higher feeding rate.

Table 13: Initial and final mean weight per fish and stocking density, as well as total weight gain, total feed input, and the feed conversion ratio (FCR) per fish charge of African catfish (*Clarias gariepinus*) production.

Number of fish charge[1]	Initial mean weight (g fish^{-1})	Initial stocking density (kg m^{-3})	Final mean weight (kg fish^{-1})	Final stocking density (kg m^{-3})	Total weight gain (kg tank^{-1})	Total feed (kg)	FCR (kg kg^{-1})
01-1-1	368.9	28.7	1.4	101.9	131.9	121.5	0.92
02-1-2	265.5	22.9	1.3	107.4	152.1	135.3	0.89
03-1-3	246.8	21.7	1.6	140.8	214.4	192.6	0.90
04-1-4	367.4	36.9	2.2	185.5	267.5	235.2	0.88
05-2-1	389.6	34.8	1.1	95.9	110.0	105.0	0.96
06-2-2	292.5	26.0	1.2	85.0	106.2	98.4	0.93
07-2-3	288.0	32.8	1.2	103.3	126.9	119.0	0.94
08-2-4	170.6	17.1	1.1	110.7	168.6	137.2	0.81
09-3-1	275.7	21.1	1.3	101.1	143.9	125.0	0.87
10-3-2	417.4	29.9	1.6	97.6	121.7	117.4	0.96
11*-3-3	620.5	53.1			123.8	113.3	-
12*-3-4	543.3	40.8			53.9	45.8	-
13*-4-1	80.4	5.6			10.0	7.5	-
Total					1730.8	1557.3	
Mean	332.8 ± 144.0	28.6 ± 11.8	1.4 ± 0.3	112.9 ± 33.9			0.91** ± 0.05

[1]Composition: consecutive fish charge number-production cycle-fish tank.

*These fish charges were not slaughtered in 2016. Therefore no FCR (kg feed input per kg weight gain) was calculated to prevent falsification.

**The declared FCR was calculated as mean FCR of fish charge 001 to 010.

Tomato plants exposed to artificial fertilisation as control and supplemented fish waste water derived by aquaponics did not show significant differences in fruit yield (Table 14). These results confirmed the investigations of Suhl et al. (2016) validating that no significant differences exist comparing DRAPS and conventional hydroponic tomato production. The revealed tomato fruit yields for aquaponics and conventional separate hydroponics were 23.7 kg m^{-2} and 26.5 kg m^{-2}, respectively, and were comparable to those gained by Suhl et al. (2016) who produced Nile tilapia and tomatoes in DRAPS in comparison to separate hydroponics. Furthermore, the gained fruit yield can even easily compete with recent separate research hydroponic systems, producing around 18 kg tomato fruit per m^2 from March or February to October, respectively, using the tomato variety Pannovy (Dannehl et al., 2014a; Dannehl et al., 2013).

Table 14: Mean tomato fruit yield, fertiliser supply, water consumption and the calculated fertiliser use efficiency of control and aquaponics during the whole experiment.

Mean total values	Control	Aquaponics
Mean marketable tomato fruit yield (kg plant^{-1})	11.5 ± 1.6^{a}	10.3 ± 0.5^{a}
Mean mineral fertiliser supply (kg plant^{-1})	0.76 ± 0.06^{b}	0.50 ± 0.04^{a}
Mean water consumption (L plant^{-1})	315.3 ± 35.9^{a}	287.5 ± 19.6^{a}
Fertiliser use efficiency (kg kg^{-1})*	15.0 ± 0.97^{a}	20.6 ± 0.57^{b}

The mean data (± standard deviation) were compared using t-test and different superscript letters indicate significant differences ($p < 0.05$).
*Fertiliser use efficiency was calculated in kg produced tomatoes per kg fertiliser addition.

6.4.2 Nitrogen concentrations in fish water (RAS water) and fish waste water

During the start-up period of the RAS, the N_{total} concentration measured in the fish water recirculating in the RAS (RAS water) rose with increasing stocking density/growth of African catfish and contained a mean of 127.6 mg N_{total} L^{-1}. After this period, the mean N_{total} concentration in RAS water increased significantly up to 246.2 mg L^{-1} (CW 21–29, using original DRAPS) and 235.1 mg L^{-1} (CW 35–39, using modified DRAPS), respectively (Figure 15A). In detail, the mean NO_3-N increased significantly from 77.3

mg L^{-1} during the start-up phase by up to 83% in the following weeks, whereas the following post-hoc test (Dunn-Bonferroni) did not show a significant difference between the first and third period. The mean NH_4-N concentration increased also significantly by 107.5% (CW 21–29) and 85.5% (CW 35–39), respectively, compared to the first period (50.3 mg NH_4-N L^{-1}). With more than 57%, the main N compound in RAS water was always NO_3^- that did not reach harmful concentrations for African catfish as described by Schram et al. (2014). However, it has to be noticed, that the NH_4-N concentration in the RAS water was relatively high (Figure 15A). Particularly, unionised ammonia (NH_3) is highly toxic for fish and must be avoided. For African catfish, a maximum of 0.34 mg NH_3-N L^{-1} is recommended (Schram et al., 2010). The relative concentrations of NH_3 and ionised ammonium (NH_4^+) depend on the pH value. Due to the relatively low pH value in the RAS water (Figure 15A), the NH_4-N concentration did not result in harmful NH_3 concentrations (Timmons et al., 2010). However, when pH values would increase up to around 7.5, as generally recommended for good growth of fishes and of nitrifying bacteria (Antoniou et al., 1990), the high NH_4-N concentration might result in toxic NH_3-N levels.

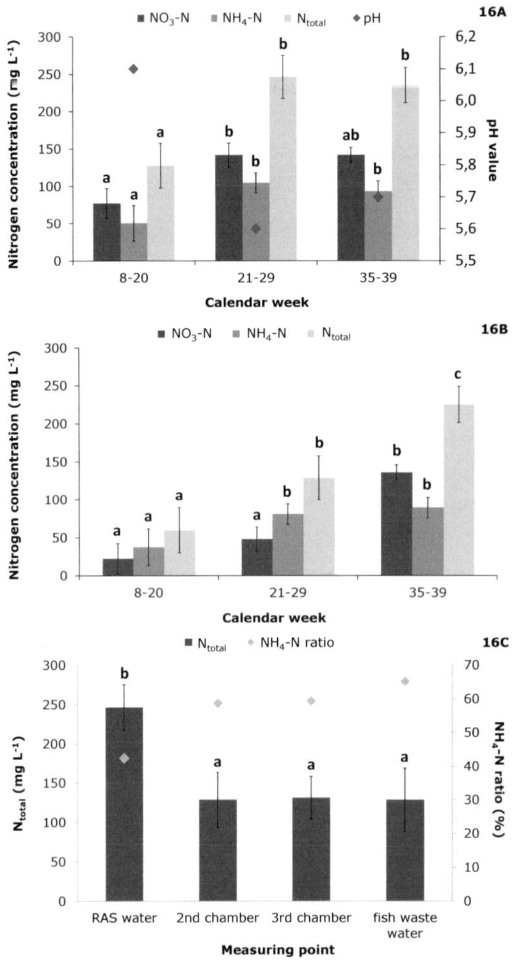

Figure 15: Nitrogen concentration and pH value in RAS water (A), fish waste water (B), and the nitrogen loss due to the 3-chamber pit measured in calendar weeks (CW) 21–29 (C). The data represent the mean concentration of nitrogen ± standard deviation of five (CW 8–20 and 21–29, and graph C) and three (CW 35–39) sampling dates (A+B). The water samples were taken in the recirculating aquaculture system (RAS water; A, B, C[11]), from the storage tank in the greenhouse (fish waste water; A, B, C[12]), and from the 2nd and 3rd chamber of the 3-chamber pit (C). The NH_4-N ratio describes the ratio of NH_4-N to total N. The different nitrogen forms were tested using post hoc tests Tukey-Kramer, Dunnett-T3 or Dunn-Bonferroni, respectively, for each kind of water between the different weeks (A+B). Different superscript letters indicate significant differences ($p < 0.05$). Abbreviations: NO_3-N = nitrate nitrogen; NH_4-N = ammonium nitrogen; N_{total} = sum of NO_3-N and NH_4-N; NH_4-N ratio = Percentage NH_4-N of N_{total}.

[11] Taken from original publication; correct would be only A

[12] Taken from original publication; correct would be only B

6.4.3 Nitrogen losses caused by the three chamber pit and its suppression

The mean N_{total} concentration in the fish waste water was only 59.8 mg L^{-1} (CW 8–20) and 128.7 mg L^{-1} (CW 21–29) (Figure 15B) and therefore, reduced by 53.1% and 47.7%, respectively, compared to RAS water (Figure 15A+B). Because the 3-chamber pit is the only component between the RAS and the storage tank, the results suggest that the N-losses occurred when the fish waste water passed the 3-chamber pit. This could be confirmed by the analysis of the nitrogen concentration within the 3-chamber pit (Figure 15C). In comparison to the RAS water, the fish waste water in the 2nd chamber of the 3-chamber pit contained significantly less N_{total} (-47.7%). The differences remained unaffected when it was further transferred to the storage tank. The N loss was obviously a result of denitrification processes occuring within the 1st chamber. The accumulated sludge and low oxygen concentration in the 1st chamber provide optimal conditions for this process in which the nitrate is converted to molecular nitrogen. Beside the loss of N_{total}, the quality of the fish waste water changed, as well. The rate of NH_4-N (percentage NH_4-N of N_{total}) was 42.4% in the RAS water, but increased significantly after passing the 3-chamber pit to 65.1% in the fish waste water (Figure 15C). Actually, a more pronounced decrease in NH_4-N concentration, caused by nitrification in the 2nd chamber of the 3-chamber pit, was desired in comparison to the observed results. Two reasons might prevent nitrification processes in the 2nd chamber. Firstly, the pH value of the fish water was below 6 (Figure 15A). But nitrification is most effective at pH levels between 7.5 and 9.0 (Hochheimer & Wheaton, 1998). Secondly, no filter material for settling of microorganisms was filled in the 2nd chamber. It is well known that the filter material and its specific surface is an important factor for adequate density of nitrifying bacteria for high nitrification rate (Krüner & Rosenthal, 1983; Lekang & Kleppe, 2000; Timmons et al., 2010).

With regard to the mentioned nitrification and denitrification phenomena revealing various nitrogen sources in the RAS, the major inorganic nitrogen sources for plants are NO_3-N and NH_4-N (Hawkesford et al., 2012). Adams (1999) recommended a maximum of 10% NH_4-N of the total N for tomatoes. While 10% NH_4-N promote growth and yield of tomatoes, concentrations ranging between 20% and 50% NH_4-N can result in lower yields of marketable fruit (Adams, 1999; Siddiqi et al., 2002). The ratio of NH_4-N to N_{total} in the fish waste water was much higher than recommended for tomatoes. Especially during CW 21 to 29, the NH_4-N concentration varied between 56.3 mg L^{-1} and 104.6 mg L^{-1} and the N_{total} concentration between 71.9 mg L^{-1} and 174.9 mg L^{-1}. This resulted in a mean NH_4-N to N_{total} ratio of 65.1% (Figure 15C). However, due to the loss of total N while passing the 3-chamber pit, nitrate fertiliser was added subsequently to the fish waste water to a final N_{total} concentration of 241 mg L^{-1} before it was delivered to the plants. This was done until CW 29. Although, the addition of nitrate fertiliser is a burden to the environment, which will be discussed in detail in paragraph 6.4.5. However, based on the addition of mineral fertiliser, the ratio of NH_4-N (56.3 mg L^{-1} to 205.6 mg L^{-1}) to N_{total} (241 mg L^{-1}) was reduced from around 65% to an average of 33%, but without affecting yields.

To prevent the N-loss in the 3-chamber pit and to reduce consequently the nitrogen fertilization, a new filter device was implemented into the RAS (modified DRAPS) to replace the 3-chamber pit, whereby the fish waste water was transferred directly into the storage tank in the greenhouse (see 6.3.1, Figure 13B). Noteworthy is that this system prevented total N losses, which was a maximum of 4.3% (235.1 mg L^{-1} in fish water to 225.1 mg L^{-1} in fish waste water) using the new filter device (Figure 15A+B). In consequence, the N concentration in the fish waste water was already as high as recommended for tomatoes (De Kreij et al., 1999) and therefore nearly no nitrogen addition was required. The proportion of NO_3-N and NH_4-N to N_{total} were 60.5% and 39.5%, respectively. Although some authors reported that high NH_4^+ concentrations can have a negative effect of tomato performance, as mentioned above, Claussen (2002) found beneficial effects of NH_4^+ on tomatoes up to an NO_3:NH_4 of 50:50, respectively. This would mean that the new filter system of the modified DRAPS using African catfish in the RAS part has the potential to provide optimal nitrogen conditions for growing tomatoes. However, a shift to higher pH values of about 7.0 in the RAS would avoid the high ammonium levels by higher nitrification rates in the biofilter. This would be advantageous for the hydroponic unit in terms of lowering the NH_4-N to N_{total} ratio.

6.4.4 *Water consumption, fresh water use efficiencies, and production ratio*

From the initial fish stocking and planting of the tomato plants until the end of the harvesting period of tomatoes (CW 42), the total fresh water supply for the RAS was 135.5 m^3. Until the end of fish production (CW 52), further 34.8 m^3 of fresh water were supplied. The total fish waste water consumption by the hydroponic unit (72 plants), was 20.7 m^3 (287.5 L $plant^{-1}$; Table 14). Regarding that and considering the total production of 1730.8 kg African catfish and a tomato fruit yield of 10.3 kg $plant^{-1}$ the fresh water use efficiency for the aquaponics treatment was 35.8 kg tomato fruit plus 10.2 kg African catfish per m^3 fresh water supply. The fresh water consumption of control plants was at a similar level (22.7 m^3, 315.3 L $plant^{-1}$) and resulted in a fresh water use efficiency of 36.5 kg tomato fruit per m^3 fresh water supply. The African catfish productivity per unit fresh water input (10.2 kg m^{-3}) can be assessed as good and was higher than reported by Food and Agriculture Organization of the United Nations (FAO, 2010) (6.6 kg m^{-3}). However, the African catfish in the present study was produced at a relative low stocking density. Although, the stocking density was below its potential, a higher density would probably not increase the productivity per fresh water input significantly. That assumption is mainly based on the fact, that higher stocking densities are coupled with a higher feed input, which consequently would result in a higher water exchange rate in order to transfer fish water at the optimum N concentration for the tomatoes immediately avoiding any higher N accumulation in the RAS.

Since only fish waste water was used to produce tomatoes using aquaponics, it was calculated that 661.3 L m^{-2} of fresh water was saved taking into account a plant density of 2.3 plants m^{-2}. Under consideration of the dimensions of the used fish rearing and greenhouse facilities in the present study, only a small amount of the available fish waste water (15.3% until CW 42) was consumed by tomato plants. While plants on a net-acreage of around 31 m^2 were provided with fish waste water containing nutrient solution, the total volume of the collected Fish waste water (135.5 m^3) would be enough for suppling plants on a net-acreage of 203 m^2. Under consideration that 135.5 m^3 fish waste water was provided by 7.2 m^3 net fish production volume, the optimal fish to plant ratio under the present conditions would have been 1 m^3 net rearing volume for African catfish to 28.2 m^2 net-acreage for tomato production. In comparison to the results of

Kloas et al. (2015), the fish and tomato production were increased by nearly 100% and 43%, respectively. While they could produce 5 kg tilapia and 25 kg tomatoes with 1 m^3 fresh water supply, 10.2 kg African catfish and 35.8 kg tomatoes were produced with the same amount of water in the present study. The higher fish production was caused by the fish species because African catfish was stocked at higher densities than tilapia. In addition, the rearing period in the present study was longer than in the mentioned study, which increased also the total output of the system. However, the generated values were quite good in comparison to single production of tomatoes or African catfish. A water use efficiency of 26.7 kg m^{-3} was reported for tomatoes grown in hydroponics with rock wool as growing medium (Dannehl et al., 2014c). In case of fish production in RAS, Verdegem et al. (2006) reported water use efficiencies ranging from 1.1 to 11.8 kg m^{-3} for catfish, eel and turbot. Extrapolated to 203 m^2 (net-acreage) a total yield of around 4.8 t of tomatoes could have been produced. Under consideration of the optimal area ratio and a total production of 1.7 t African catfish the production ratio (kg fish production per kg tomato fruit yield) has been extrapolated to be 1 kg fish to 2.8 kg tomatoes.

6.4.5 *Fertilizer savings and fertilizer use efficiency and causing environmental relief*

The monthly mineral fertiliser addition varied from 1.42 to 2.89 g L^{-1} and from 0.35 to 2.09 g L^{-1} in order to mix the nutrient solution for control plants or for plants exposed to the fish waste water based nutrient solution, respectively (Table 15). From June onwards the fertiliser reduction (39.1%) caused by fish waste water was significant compared to the control. The fertiliser reduction ranged between 13.2% and 77.7% within seven months.

After modifying the DRAPS (August, September), more than 50% fertiliser was saved and the highest reduction was realised in September with 77.7%. Le Bot et al. (2001) reported that tomatoes produced under hydroponic conditions receive around 2000 kg N ha^{-1}. Wood and Cowie (2004) described that in Germany for a mean N fertiliser manufacturing a CO_2 equivalent ($CO_{2\text{-}e}$) of 5.3–7.6 kg per kg N is released into the environment. In this context, since nearly 100% nitrogen was provided by fish waste water exposed to the modified DRAPS, a total amount of 10.6–15.2 t $CO_{2\text{-}e}$ per ha (1.1–1.5 kg $CO_{2\text{-}e}$ m^{-2}) tomato production in greenhouses can be saved when the N demand of 2000 kg N ha^{-1} is considered for calculations. Furthermore, it has to be kept in mind that an adjustment of the nutrient stock solution depending on the current nutrient concentrations in the fish waste water is absolutely crucial for a well performing DRAPS, whereby a maximum of fertiliser saving is possible. However, under consideration of the total marketable yield per plant the lower fertiliser use (13.2–77.7%), mentioned above, revealing a significantly increased fertiliser use efficiency (37.3%) in aquaponics (Table 14). While the application of 1 kg fertiliser to control plants resulted in a production of 15.0 kg tomato fruit, 20.6 kg tomatoes were produced in aquaponics with the same amount of fertiliser supplementation. However, Suhl et al. (2016) reported a higher total fertiliser use efficiency for DRAPS produced tilapia and tomatoes, which was more than double as high as presented in the current study. This was mainly based on the fact that less fertiliser (EC 1.8 dS m^{-1}) was used by the mentioned authors, where approximately the same tomato fruit yield was achieved. In the present study, however, an EC value of 2.6 dS m^{-1} was used, especially to increase fruit quantity and quality as described in different studies (Dorai et al., 2001; Sonneveld & Welles, 1988). Considering the results of the present study, it can be concluded that a higher EC value under present conditions is not necessary. Additionally, the fertiliser use efficiency in the present study was calculated for the whole production period. Considering that the fertiliser reduction increased markedly over time (August, September; Table 15) a much higher fertiliser use efficiency for aquaponics can be expected and could reach up to 77.7% based on the data for September (Table 15).

Table 15: Effect of time course and different process waters on fertiliser addition1.

Month	Fertiliser addition (g per L) Control	Aquaponics	Fertiliser reduction* (%)
March	1.42 ± 0.36^a	1.23 ± 0.22^a	13.2
April	2.39 ± 0.04^a	1.86 ± 0.27^a	22.0
May	2.54 ± 0.24^a	2.09 ± 0.11^a	17.8
June	2.89 ± 0.20^b	1.76 ± 0.15^a	39.1
July	1.75 ± 0.19^b	1.34 ± 0.14^a	23.6
August	1.70 ± 0.25^b	0.83 ± 0.06^a	51.2
September	1.57 ± 0.11^b	0.35 ± 0.03^a	77.7

The data represent mean values (± standard deviation) of added total mineral fertilizer per L nutrient solution and were compared using t-test and Mann-Whitney *U*-test. Significant differences are indicated by different small letters ($p < 0.05$).
[1]Total fertiliser addition: calcium ammonium nitrate (containing 1.1% ammonium), potassium nitrate, magnesium sulphate, potassium hydrogen phosphate, and ready to use trace elements solution.
[*]Fertiliser reduction in aquaponics compared to control (= 100%).

6.5 Conclusion

The modification of the already existing double recirculating aquaponic systems (DRAPS) by the implementation of a new suction filter device successfully prevented nitrogen loss and increased the fertiliser saving potential. As such, the monthly mean fertiliser input to the conventional hydroponic unit was reduced up to 39.1% using the original DRAPS. After modification to reduce nitrogen loss caused by the implemented 3-chamber pit, the fertiliser input was even reduced by 77.7%, compared to the control. A reduced fertiliser usage in turn can reduce the environmental emissions. Because the waste water of African catfish can deliver nearly 100% of the nitrogen required by tomatoes, the modified DRAPS can reduce the impact to global warming enormously alone by reducing greenhouse gas emissions during the manufacturing of nitrogen fertiliser that would be needed to grow tomatoes in separated hydroponic systems.

The present study shows for the first time that also African catfish and tomato can be produced in DRAPS in the same quantity as in separate fish and tomato production sites.

However, it was found that waste water from African catfish, reared as described, contains high concentrations of ammonium. Despite in the present study no negative effects on fruit quantity were found, further investigations are warranted to answer whether the direct use of the fish waste water derived by such modified DRAPS could cause adverse impacts on tomato fruit quantity and quality on a long term basis.

7

Prospects and challenges of double recirculating aquaponic systems (DRAPS) for intensive plant production

Johanna Suhl, Dennis Dannehl, Leonard Zechmeister, Werner Kloas, Benjamin Lehmann, Günther Scheibe, and Uwe Schmidt

Acta Horticulturae 1227 (2018) 449–456
DOI: 10.17660/ActaHortic.2018.1227.56

7 Prospects and challenges of double recirculating aquaponic systems (DRAPS) for intensive plant production

7.1 Abstract

In contrast to conventional single recirculation aquaponic systems (SRAPS), the innovative double recirculating aquaponic system (DRAPS) provides an opportunity for intensive fish and plant production. This system is based on two separate recirculating systems, a recirculating aquaculture system (RAS) and a closed hydroponic system, whereby the composition of fish waste water can be optimized for plant growth by the addition of synthetic fertilizer. However, in the last two years of research it has become apparent that the first approach to DRAPS needs to be reconsidered and further optimized. In the first implementation of DRAPS, the fish waste water passed a sedimentation filter incorporated in the RAS and then a secondary clarifier, constructed as a three-chamber pit, for deposition of solids before delivery to the hydroponic system. It became clear that the depletion of nitrogen caused by the secondary clarifier was too high. Therefore, a new approach was developed and is addressed in the present study: a new suction filter was developed for sludge elimination in the RAS. The first weeks of its use were promising and it is obviously possible to omit the previously used sedimentation filter and leave out the secondary clarifier completely. With this setup, the fish waste water from the RAS was used directly with a minimum of water losses. Another advantage was that the nitrogen concentration in the RAS was kept relatively stable by the use of the new suction filter.

7.2 Introduction

Aquaponics combines fish production in recirculating aquaculture systems (RAS) and plant production in soilless hydroponic culture. Water and nutrients can be used twice, firstly for fish rearing and secondly for plant irrigation, enabling sustainable food production. The savings of mineral fertilizer, for instance nitrogen fertilizer, can reduce the environment pollution. During the manufacture of nitrogen fertilizer, large amounts of CO_2 and different oxides of nitrogen (Wood & Cowie, 2004) are emitted, which have a major impact on global warming or eutrophication (Antón et al., 2004; Torrellas et al., 2012). Conventional aquaponics work as low-tech single recirculating aquaponic systems (SRAPS). In such systems, the nutrient-enriched fish water circulates between fish-rearing tanks and the hydroponic unit (Diver & Rinehart, 2010). In this case the plants and the microorganisms living in the rooting zone of the plants are used to eliminate nutrients from the fish water (Rakocy et al., 2006; Tyson et al., 2011). Because of the differing environmental requirements of fish, plants and microorganisms, a compromise regarding the water quality must be found and usually results in yields below their potential (Graber & Junge, 2009; Vergote & Vermeulen, 2012; Wortman, 2015). Consequently, SRAPS are not used for intensive crop production, but mostly for hobby or for education (Love et al., 2014). To use aquaponics as sustainable, but intensive fish and crop production system, an innovative double recirculation aquaponic system (DRAPS) were developed by Kloas et al. (2015). DRAPS also known as decoupled aquaponics, allows the intensive production of fish and plants (Kloas et al., 2015; Suhl et al., 2016) in two separated cycles which operating independently. The separation of the units allows the adjustment of optimal conditions such as pH and nutrient composition for both species independently. The structural construction of the DRAPS prototype, developed and published by Kloas et al. (2015), were modified and described by Suhl et al. (2016). Briefly, a secondary clarifier, a spatial separation of the fish and plant production unit, a reduced fish to plant ratio, and a continuous adjustment of the nutrients solution mixed for the plants were the main modifications. The addition of the secondary clarifier was intended mainly to reduce water wastage. In this setup, the mechanical filter in the RAS must be drained and cleaned regularly to remove sludge from the RAS and to keep the nutrient, especially nitrogen, content low by fresh water replacement. The drained water is released into the secondary clarifier to separate the sludge and to convert ammonium to nitrate. The fish waste water is then usable for plant irrigation.

During the operation of the new DRAPS it became clear that the secondary clarifier caused loss of nitrogen. Because this is not compatible with sustainable aquaponics, the present study focused on the implementation of a new device to replace the mechanical filter and consequently the secondary clarifier without increasing the water demand of the whole system. The present study describes the innovative and newly developed suction filter system to remove sludge from the RAS directly and its connection into the DRAPS described by Suhl et al. (2016) and resulting modifications. For evaluation of the new suction filter, the nitrogen content of the fish water and its variation through the day and from day to day were investigated. To that end, mineral nitrogen was measured hourly and daily.

7.3 Materials and methods

7.3.1 System design

The experiments were carried out in a research DRAPS consisting of a 139 m² Venlo-type greenhouse, a 43 m² fish farm and a technical room with 14 m² as described exactly by Suhl et al. (2016). The schematic diagram of the original DRAPS is described and pictured by Suhl et al. (2016).

The fish were reared in four identical glass-fibre tanks with a net production volume of 1.8 m^3 each. The water flowed from the fish tank into the mechanical sedimentation filter (1.3 m^3) and then into a pump sump. From there it was pumped into a trickling biofilter for nitrification – the biological oxidation of ammonium (NH_4^+) to nitrite (NO_2^-) followed by the oxidation of nitrite to nitrate (NO_3^-). After passing the biofilter, the fish water flowed through a reception tank back into the fish tanks. To remove sludge and nutrients from the original DRAPS (Suhl et al., 2016), the mechanical filter was cleaned regularly, depending on fish water quality. To clean the mechanical filter, the water was released into the secondary clarifier. The secondary clarifier itself consisted of three chambers, the first unaerated to sediment the sludge, the second aerated to prevent denitrification [biological reduction of NO_3^- to nitrogen gas (N_2)] and the third (unaerated) to pump the fish water into the greenhouse. The fish water was collected there in a storage tank until being used in mixing the nutrient solution. To mix the nutrient solution, the fish waste water was enriched with missing minerals using synthetic fertilizer.

7.3.2 Fish production

In 2015, before the original DRAPS were modified, Nile tilapia (*Oreochromis niloticus*) were reared at a low stocking density. The exact key production figures were described in detail by Suhl et al. (2016). In 2016, the rearing of African catfish (*Clarias gariepinus*) began, and this continued in 2017. After a start-up phase, the African catfish were stocked at a mean density of around 75 kg m^{-3}, and the total mean feed conversion ratio was 0.89.

7.3.3 *Nitrogen loss due to the secondary clarifier while using the original DRAPS*

To investigate the amount of nitrogen lost in the secondary clarifier under the original DRAPS, water samples were taken from the reception tank of the RAS and from the storage tank in the greenhouse. The samples were taken once a week for ten consecutive weeks in two periods (March until May 2015 and September until November 2015, tilapia rearing). For representation, the measured concentration for each measuring period was averaged. Until analysis the samples were stored at -20°C. Nitrate nitrogen (NO_3-N) and ammonium nitrogen (NH_4-N) were analysed using continuous flow analysis (San++, Fa. SKALAR; Breda, The Netherlands) and the measuring method is exactly described by Suhl et al. (2016). The concentration of NO_3-N and NH_4-N were summarized to mineral nitrogen (N_{min}) and expressed as mg L^{-1}.

7.3.4 *The new suction filter*

A new and innovative suction filter was developed for purifying the recirculating RAS water (PAL KLAR; Patent number DE 102014004767A1 (Lehmann et al., 2015)). Two prototypes were constructed. Prototype II, finished in 2017, was designed to improve some technical aspects of prototype I (finished in 2016). In general, both worked in the same way. The suction filter consists of a combination of a sieve and a suction device. The circular stainless-steel sieve circulates in a separate water tank placed within the RAS between the fish-rearing tanks and the pump sump (Figure 16). Between in- and outflow the sieve circulates and sludge adheres to it. The sieve is installed half above and half below the water surface. Above the water surface the tube of the suction device is controlled by the water level and moves on a linear glide rail in front of the sieve. The suction flow removes solids from the mesh. In addition, prototype II contains a backwash function. The backwash cleans the sieve, and also the suction tube, regularly. The daily freshwater consumption of the DRAPS can be controlled by the flat film extrusion in the backwash apparatus, which can be programmed using a control device. It is even possible to suspend the backwash for few days. The different key figures of both prototypes are listed in Table 16.

Table 16: Key features of the two prototypes of the new suction filter unit in the recirculating aquaculture system.

	Prototype I	Prototype II
Mode of operation of the suction filter[1]	Continuous	Discontinuous
Connected wattage of the suction filter	4.5 KW	4.5 KW
Power consumption		1.3 KW
Sieve mesh width	150 µm	150 µm
Tank volume	700 L	370 L
Other		Backwashing unit for sieve

[1]Continuous, the suction tube swings continuous for 24 h a day along the sieve; discontinuous, the suction tube swings discontinuous along the sieve controlled by water level.

7.3.5 Modifying the double recirculating aquaponic system

The new suction filter replaced the mechanical filter in the RAS system (Figure 16). The sludge in the RAS was removed by the new suction filter (see above) and made the secondary clarifier unnecessary. However, to maintain water quality and to keep the nitrogen level in the RAS at an adequate level, a water exchange in the RAS was still necessary. This water exchange occurred according to the withdrawal of fish water used for plant irrigation. Therefore, a pump transferred the fish water five times a day, around 40 L each time, from the reception tank of the RAS into the storage tank in the greenhouse. The operation of the suction filter caused additional water loss because of the backwashing function of prototype II.

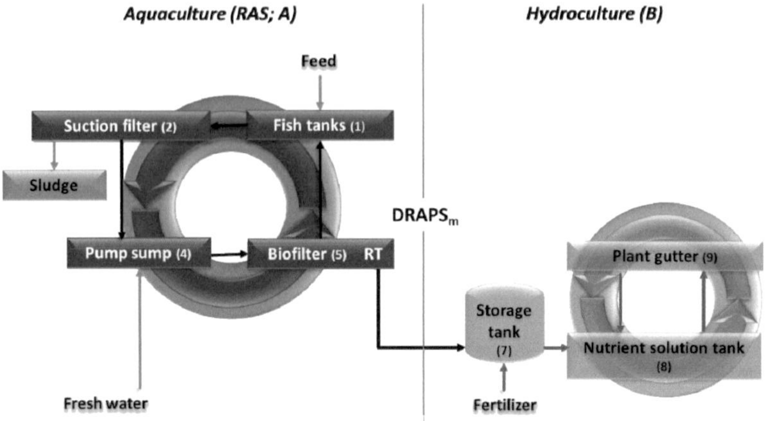

Figure 16: Schematic diagram of the modified double recirculating aquaponic system (DRAPS). (A) Recirculating aquaculture system (RAS) with fish-rearing tanks (1), suction filter (2), pump system (3), pump sump (4), and trickling biofilter (5). (B) Recirculating hydroponic unit with storage tank for fish water (7), nutrient solution tanks (8) and plant gutters (9). Both systems are connected via a pipe through which water is pumped from the reception tank (RT) to the storage tank in the hydroponic unit.

7.3.6 Nitrogen content in fish water while using the new suction filter

To determine the nitrogen concentration in the fish water while using the modified DRAPS with the new suction filter on an hourly and daily basis, samples were taken in two different periods. The nitrogen variability during the day were tested in 2016, when the prototype I were implemented in the RAS (African catfish rearing). Therefore, the samples were taken every full hour between 8 a.m. and 6 p.m. This occurred on three consecutive days at the end of September 2016. The following investigation, to evaluate the course of the nitrogen concentration on 15 consecutive days, occurred in Mai 2017. At this time the prototype II were implemented into the RAS and the samples were taken every day at 9 a.m. from 17.05.2017 until 30.05.2017. The samples for both investigations were taken from the reception tank of the RAS and from the storage tank in the greenhouse (Figure 16) to evaluate the nitrogen loss and nitrogen concentration. Until analysis the samples were frosted and stored at -20°C. Nitrate nitrogen (NO_3-N) and ammonium nitrogen (NH_4-N) were analysed using ready to use cuvette tests (LCK 303 and LCK 340, Hach Lange GmbH; Düsseldorf, Germany). The analysis followed the instruction given by manufacturer. Fore analysis the fish water were diluted with distilled water by dilution factor five for NO_3-N and three for NH_4-N. For representation in the present study the NO_3-N and NH_4-N concentration were summarized and expressed as N_{min} in mg L^{-1}.

7.3.7 Statistical analysis

To evaluate the nitrogen loss by the secondary clarifier the nitrogen concentration in the fish water and in the storage tank were tested using t-test ($p < 0.05$) after normal distribution was confirmed. Standard deviations are displayed as barks and significant differences are indicated by small letters. For statistical analysis SPSS package version 19.0 was used.

7.4 Results and discussion

7.4.1 Nitrogen loss due to the secondary clarifier

It was shown that the fish water lost significant amounts of nitrogen while it passed through the secondary clarifier (Figure 17). In both measuring periods nearly 45% and 30% of N_{min} got lost, respectively. In the first measuring period (March until May 2015) the N_{min} concentration decreased from 45.0 mg L^{-1} to 23.3 mg L^{-1} and in the second period from 105.8 mg L^{-1} to 74.2 mg L^{-1}. The loss of nitrogen may be the result of denitrification processes occurred in the first chamber of the secondary clarifier. The sludge sedimented to the ground and was presumably an excellent environment for anaerobic denitrification bacteria.

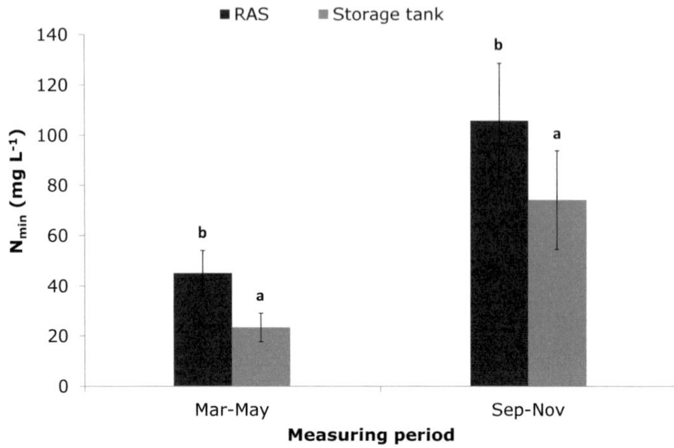

Figure 17: Nitrogen concentration (N_{min} = NO_3-N + NH_4-N) in Nile tilapia water, taken from the reception tank of the recirculating aquaculture system (RAS) and in the fish water stored in the storage tank in the greenhouse after it passed the secondary clarifier. The data represent mean values of respectively ten consecutive weeks in two measuring periods (March – May and September – November), respectively. Mean values were tested using t-test ($p < 0.05$) and small letter indicate significant differences.

7.4.2 Effect of the new suction filter on nitrogen content

The implementation of the new suction filter resulted in a consistent water exchange. As reported by Suhl et al. (2016), the original mechanical filter was cleaned one to three times a week. With each cleaning operation, a total volume of 1.3 m^3 was removed from the RAS. With the new suction filter a daily amount of around 200 (prototype I) to 330 L (prototype II, 200 L for plants and 130 L for filter operation) was exchanged. The consistent water exchange resulted in a very stable nitrogen concentration through the day (Figure 18). The nitrogen analysis showed that the nitrogen content during the day did not fluctuate strongly. Even over three consecutive days of investigation, nitrogen content was relatively stable and fluctuated only between 253 and 283 mg L^{-1}. A constant nitrogen concentration is a key requirement in using fish water in DRAPS. Suhl et al. (2016) and Kloas et al. (2015) reported a clearly greater nitrogen fluctuation in a RAS used in a DRAPS. To deal with this strong fluctuation, Suhl et al. (2016) recommended an automatic online measuring system with direct feedback control for reliable and constant adjustment of the fish water to optimise conditions for plant growth. Even if an automatic adjustment system is an effective solution for DRAPS, it implies greater technical and economic investment. With a stable nutrient concentration, the adjustment of the fish water is much easier and an automatic system can be dispensed with and costs saved.

Furthermore, a much smaller nitrogen loss was detected, because of the elimination of the secondary clarifier. On average, around 17.6 mg L^{-1} less nitrogen was measured in the storage tank than in the RAS. At a mean nitrogen concentration of 268 mg N_{min} L^{-1} in the RAS water, the reduction was only around 6.6%, compared with 30–45% with the use of the secondary clarifier (see above).

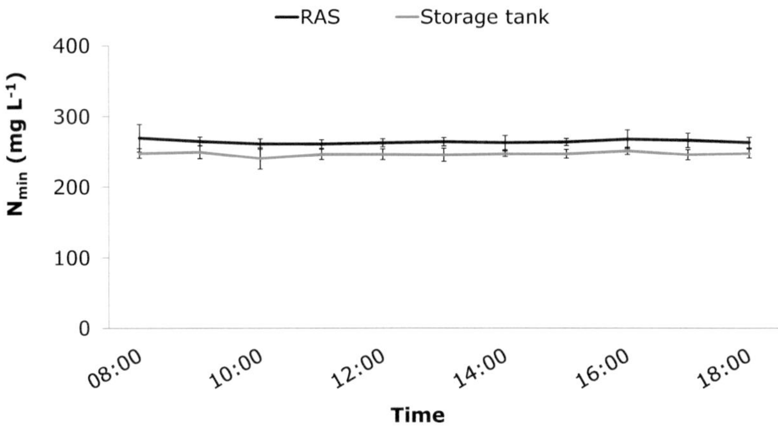

Figure 18: Nitrogen concentration (N_{min} = NO_3-N + NH_4-N) in African catfish water, taken from the reception tank of the recirculating aquaculture system (RAS) and the storage tank in the greenhouse. The data represent mean values of three consecutive days measured in September 2016. The data represent mean values ± standard deviation.

After determining in 2016 that nitrogen content was stable during the day and for three days in a row (Figure 18), the nitrogen balance over 2 weeks was analysed again in 2017, after some technical improvements to the construction and the development of prototype II. It was shown that, measured on a daily basis for 15 consecutive days, the nitrogen content was less stable (Figure 19) than in the first investigation in September 2016 (prototype I, Figure 18). A slow increase in nitrogen content over the 15 days was detected. The concentration increased from 163 mg L^{-1} on the first measuring day by 100 to 263 mg L^{-1} on the last measuring day. In the middle of the measuring period (23 May), a large decrease in nitrogen concentration to 122.5 mg L^{-1} was detected. The drop was detected for one day only and cannot easily be explained. The drop may have been caused by a measuring error, since the average nitrogen concentration recorded afterwards was as high as the day before the drop.

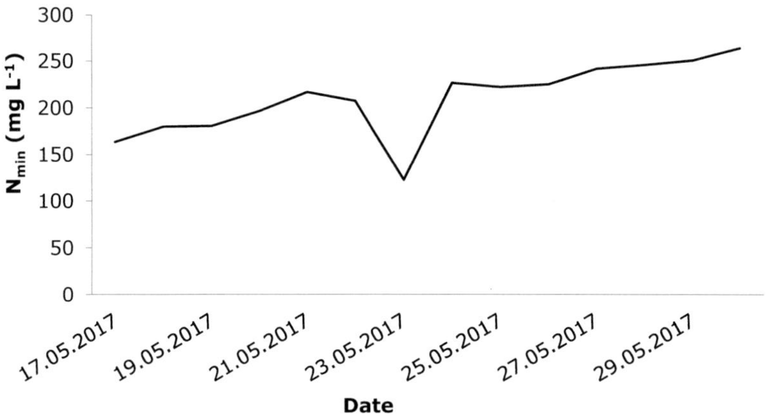

Figure 19: Nitrogen concentration (N_{min} = NO_3-N + NH_4-N) in African catfish water. The samples were taken in Mai 2017 from the reception tank of the RAS on 15 consecutive days.

The slowly increasing concentration is not necessarily desirable under consideration of the adjustment of the fish water for plant irrigation as discussed above. It is to be expected, that a higher exchange rate, means a higher fish water transfer to the hydroponic unit, may reduce the accumulation of nitrogen and a stabilisation as measured in September 2016 (Figure 18). Further investigations are necessary to find the right exchange rate for a stable nitrogen concentration for several days/weeks. Additional, the increasing nitrogen content in the RAS to above 263 mg L^{-1} was too high for optimal tomato nutrition. Depending on the resource the recommended nitrogen concentration in nutrient solution is between 151 mg L^{-1} (Lattauschke, 2004a) and 241 mg L^{-1} (De Kreij et al., 1999). It was demonstrated, for instance, that increasing nitrogen supply via nutrient solution to tomatoes may increase the susceptibility for pests (Hoffland et al., 2000a) and bacterial and fungal (Hoffland et al., 2000b) infections. This must be avoided. In summary, further investigations are necessary to confirm the potential of the new suction filter device and to evaluate its benefits regarding nutrient management in DRAPS.

7.5 Conclusion

The present study demonstrated that the mechanical filter and the secondary clarifier can be successful replaced by the new developed suction filter. Additional to a more consistent water exchange in the RAS a relative consistent nitrogen concentration was detected. The very stable nitrogen concentration, measured on an hourly basis for three consecutive days, could not be confirmed in the second measuring period were the concentration were measured on a daily basis for 15 consecutive days. Nevertheless, it was shown that the loss of nitrogen could be reduced by the new suction filter device. Altogether the new suction filter was a technical enrichment of the DRAPS and has the potential to enhance the sustainability of the original DRAPS. Further tests are necessary to confirm the long-term stability of the system.

8

Lettuce (*Lactuca sativa*, variety Salanova) production in decoupled aquaponic systems: Same yield and similar quality as in conventional hydroponic systems but drastically reduced greenhouse gas emissions by saving inorganic fertilizer

Hendrik Monsees*, Johanna Suhl*, Maurice Paul, Werner Kloas, Dennis Dannehl, and Sven Würtz

*These authors contributed equally to this work.

PLoS ONE 14(6): e0218368

DOI: 10.1371/journal.pone.0218368

8 Lettuce (*Lactuca sativa*, variety Salanova) production in decoupled aquaponic systems: Same yield and similar quality as in conventional hydroponic systems but drastically reduced greenhouse gas emissions by saving inorganic fertilizer

8.1 Abstract

Decoupled aquaponic systems have the potential to become one of the most effective sustainable production systems for the combined production of animal protein and plant crops. Here, recirculating aquaculture systems for fish production are combined with hydroponics for soilless plant production thereby recycling dissolved nutrients derived from metabolism of the fish. The aim of the present study was to characterize hydroponic lettuce production using conventional nutrient solution in comparison with decoupled aquaponics using the nutrient rich fish water as basis for the nutrient solution being supplemented by missing nutrients. In addition, one aquaponic treatment became disinfected in order to assess any occurring advantage of the aquaponics derived fish water. For evaluation the temperature, electrical conductivity, pH, and the mineral composition of the nutrient solution, as well as colony forming units in the fish water were monitored. Additionally, plant growth (fresh and dry weight, number and area of leaves) and quality parameters of lettuce leaves (nitrate, mineral content, phenolic compounds) were examined. Carbon sources and microorganisms derived from fish water seem to have neither beneficial nor detrimental effects on plant growth in this study. Except for some differences in the mineral content of the lettuce leaves, all other quality parameters were not significantly different. The use of aquaponic fish water saved 62.8% mineral fertilizer and fully substituted the required water for the nutrient solution in comparison to the control. Additionally, the reduced fertilizer demand using decoupled aquaponics can contribute to reduce greenhouse gas emissions of an annual lettuce production site per ha by 72% due to saving the energy for fertilizer production. This study clearly demonstrates the huge potential of the innovative approach of decoupled aquaponics to foster the transformation of our conventional agriculture towards sustainable production systems saving resources and minimizing emissions.

8.2 Introduction

The conventional production of animal protein for human consumption produces high amounts of liquid and organic waste (urine and faeces) that is regularly used to fertilize plant crops (UBA, 2017). Due to its undirected application on agricultural landscapes this fertilization practise often results in eutrophication and pollution of the adjacent water bodies (UBA, 2017). As Tilman et al. (2011) reported, the global crop demand will increase by 100–110% from 2005 to 2050. The intensification of agriculture in turn, may have an immense negative impact on the environment (Tilman et al., 2002). As one approach to minimize the environmental impacts the so called "sustainable intensification" aims to increase the yield per unit area without increasing environmental burden (Godfray et al., 2010; Tilman et al., 2002). Sustainable intensification of crop production is hence a goal of the "Strategic Objective A" of the Food and Agriculture Organization of the United Nations (FAO). The FAO developed three intertwined pillars to reach sustainable intensification. Parts of them are the efficient use of resources, but also the protection of the environment by reducing pollution. One possibility to enforce these aims is aquaponic production systems as described below.

Within the last decade aquaculture has been the fastest growing sector in animal food production with average growth rates of 5.4% (Bostock et al., 2010; FAO, 2016a). Ongoing research and development provided the basis for a constant improvement of common aquaculture practises and increased the sustainability of aquaculture systems. In this context, recirculating aquaculture systems (RAS) have been developed to address several bottlenecks of open production systems (van Rijn, 2013). Here, aquatic animals (mainly fish) are produced in closed systems with an integrated water treatment e.g. including biofilters for the conversion of ammonium (NH_4^+) to nitrate (NO_3^-) (nitrification) and mechanical filtration units (e.g. drum filter) for effective solids removal (van Rijn, 2013). As a result of water treatment and low water consumption, nutrients (mainly nitrogen in the form of nitrate) and organic matter (often specified as dissolved organic carbon (DOC) and total organic carbon (TOC)) accumulate within these systems (Hambly et al., 2015; Martins et al., 2010; Yamin et al., 2017). Consequently, if RAS are used for fish production, a recycling of RAS-derived nutrients for plant production should be promoted to close nutrient cycles and improve the resource efficiency. Nevertheless, it would be also possible to remove the nitrate via denitrification units (van Rijn, 2013), but here nitrogen would be lost, mainly in the form of N_2.

Aquaponics is defined as the combination of fish rearing and hydroponic (soilless) plant production (Rakocy et al., 2006). Water and accumulated nutrients from the fish unit, especially nitrogen, derived from the metabolism of the fish, but also from uneaten feed and organic debris (Timmons et al., 2010), are used as fertilizer by the plants grown in hydroponics. In conventional aquaponics, also described as coupled aquaponic systems, the water flows from the fish unit to the plant unit and back (Rakocy et al., 2006; Tyson et al., 2011). As such, compromises have to be made in terms of certain water parameters (e.g. pH, temperature, nutrient concentrations) either for the fish or for the plant production sites (Monsees et al., 2017b) leading to suboptimal conditions affecting adversely productivity of both, fish and plants. In addition to the well-established coupled aquaponic systems, a decoupled approach may also be used (Kloas et al., 2015), also referred to as double recirculating aquaponics system (DRAPS) (Kloas et al., 2015; Suhl et al., 2016). Here, fish are reared in a RAS and the plants are grown in a separate hydroponic system. The fish water from the RAS is directed to the reservoir of the hydroponic system on demand but not back to the RAS as it is the case in coupled aquaponics. Thereby the water parameters for each produced species (e.g. Nile tilapia, *Oreochromis niloticus* and tomatoes, *Solanum lycopersicum*) can be optimized separately. As such, the nutrient concentrations and pH of the fish water can be adjusted to the specific demands of the growing crop. Thus, decoupled aquaponic systems maintain the benefits of aquaponics but simultaneously provide the opportunity to generate comparable productivities as in single conventional RAS and hydroponic units, respectively (Kloas et al., 2015; Suhl et al., 2016). Besides more theoretical considerations of decoupled aquaponics (Goddek et al., 2016a) or passing remarks as innovative approach (Goddek et al., 2015; Junge et al., 2017; Villarroel et al., 2016), only a few empirical studies (Delaide et al., 2016; Kloas et al., 2015; Suhl et al., 2016) are published yet. To our knowledge, only two publications comparing decoupled aquaponics to conventional hydroponics have been published so far (Delaide et al., 2016; Suhl et al., 2016). Delaide et al. (2016) reported an increased lettuce growth of 39% in decoupled aquaponics compared to single hydroponic treatment and concluded that plant promoting effects like dissolved organic matter or growth-promoting rhizobacteria and/or fungi originating from the RAS water could be responsible for such an increased plant growth. On the contrary, Suhl et al. (2016) revealed a comparable tomato production in decoupled aquaponics and single hydroponics and confirmed that the productivity for tomatoes in DRAPS can reach the same level as in single hydroponics using optimum fertilizer but did not exceed the yield further as indicated by the results of Delaide et al. (2016) concerning lettuce.

Therefore, the aim of this study was to evaluate the ability to produce butterhead lettuce (*Lactuca sativa*, variety Salanova) in already established DRAPS as effectively as conventional hydroponics. The study was done with special emphasis on potential beneficial effects of dissolved organic matter and microorganisms derived from RAS and to highlight the resource efficiency of these aquaponic systems (in terms of water and nutrients). Finally the potential value of decoupled aquaponics for a sustainable intensification of agricultural production was evaluated.

8.3 Material and Methods

8.3.1 Lettuce cultivation and experimental set-up

The experiments were conducted from 16th March 2016 until 10th May 2016. During this period green open butterhead lettuce Salanova (*Lactuca sativa* L., Rijk Zwaanm; De Lier, The Netherlands) cv. Descates RZ were grown hydroponically in a nutrient film technique system (NFT). The experiments were conducted in a 75 m^2 compartment of an experimental Venlo-type greenhouse located in Berlin, Germany (52°46´82.806´´N, 13°29´88.909´´E). The ventilation was opened above 17 °C and the target temperature for the floor level heating system was 10 °C for night and 14 °C for day. The mean climate values during the cultivation period are displayed in Table 17.

Table 17: Greenhouse climate during the experimental period of 7 weeks. The data represent the weekly mean values of temperature, relative humidity, carbon dioxide (CO_2), and global radiation. Except global radiation, the mean values during daytime and night time are shown. For global radiation only the values for daytime are displayed.

Date		**Temperature (°C)**	**Relative humidity (%)**	**Global radiation (W m^{-2})**
18.03.-	day	14.78	54.47	174.47
27.03.2016	night	13.24	67.26	
28.03.-	day	13.36	39.64	277.36
03.04.2016	night	9.26	44.58	
04.04.-	day	16.59	46.85	255.66
10.04.2016	night	13.56	62.82	
11.04.-	day	15.55	58.43	186.32
17.04.2016	night	13.55	67.05	
18.04.-	day	15.60	42.22	359.82
24.04.2016	night	13.45	56.32	
25.04.-	day	15.88	43.03	329.52
01.05.2016	night	13.29	55.45	
02.05.-	day	21.80	40.33	414.72
10.05.2016	night	16.28	53.45	

The lettuce seeds were sown in autoclaved cultivation soil where the plants grew for two weeks. After this period, the plants were transferred into rock wool cubes (7.5 cm x 7.5 cm, Cutilene; Tilburg, The Netherlands) using the cultivation soil to fix the plants and transplanted into high gullies on 18th March 2016. A total of nine gullies (6 m length x 0.35 m width x 0.05 m high) with an inclination of 1% were installed in the greenhouse compartment. Ten lettuce plants were planted per gully with a distance of 50 cm apart (Figure 20). The space between the plants was covered with black and white plastic film to avoid algae formation. The distance between the gullies was around 0.5 m. At the end of each gully, a 300 L tank was installed as a reservoir for the nutrient solution. Conventional pumps (EHEIM universal 600, EHEIM GmbH & Co. KG; Deizisau, Germany) were used to pump the nutrient solution continuously (24 h d^{-1}) to the end of each gully from where it was flowing back in the tank.

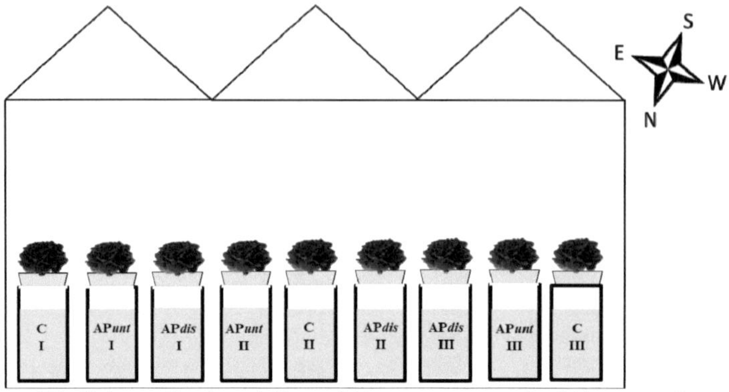

Figure 20: Schematic description of the arrangement of the experiments. C = control, fresh water based nutrient solution; AP*unt* - untreatend aquaponics, fish water based nutrient solution with supplemented nutrients; AP*dis* - like AP*unt*, but fish water was disinfected before use.

For the experiment, three different treatments were applied in triplicates (Figure 20). As such, nine gullies with ten lettuce plants each were used. In detail two nutrient solutions were prepared with fish waste water obtained from a recirculating aquaculture system rearing tilapia (see 8.3.1.2) and fertilizer addition. Afterwards, one of them was used directly (AP*unt* = Aquaponic untreated) for the experiments, whereas the other one was disinfected (AP*dis* = Aquaponic disinfected) before being used. For disinfection, the prepared nutrient solution was heated in a 150 L pot made from stainless steel to > 95 °C for 15 min using a butane gas burner (3.6 kW) and three heating rods (stainless steel, 2 kW each). Afterwards the fish water was cooled down by a plate type heat exchanger. The preparation of the nutrient solution and disinfection procedure occurred at 16th March 2016. Both treatments were compared to a control (C = control), which was prepared with fresh tap and rain water (50:50, v/v, to reduce the amount of calcium) and mineral fertilizer.

8.3.1.1 Adjustment of the nutrient solution

For optimal growth conditions, all nutrient solutions were adjusted with mineral fertilizer to an electrical conductivity (EC) of 2.2 dS m^{-1}. The optimal nutrient solution for lettuce contained 200 mg N L^{-1}, 62 mg P L^{-1}, 150 mg K L^{-1}, 210 mg Ca L^{-1}, 50 mg Mg L^{-1}, 70 mg S L^{-1}, 2.5 mg Fe L^{-1}, 0.62 mg Mn L^{-1}, 0.03 mg Mo L^{-1}, 0.09 mg Zn L^{-1}, 0.50 mg Cu L^{-1}, and 0.44 mg B L^{-1} as described by Hochmuth (2001). The mixture of rain and tap water (control) and fish water were analysed for its nutrient concentrations by inductively coupled plasma-optical emission spectrometry (ICP-OES) and continuous flow analysis (CFA) (see 8.3.2.1). Based on these results the required additions of mineral fertilizer were calculated to adjust the nutrient solution as given above. For preparing the nutrient solution, standard fertilizers were used: calcium ammonium nitrate (CAN; containing 1.1% ammonium), potassium nitrate (KNO_3), mangnesium sulfate ($MgSO_4$), potassium hydrogen phosphate (KH_2PO_4), and ready to use trace elements solution (YaraTera Tenso Cocktail, 176 Yara UK Limited; N E Lincolnshire, United Kingdom). The adjustment of the pH value was done by supplementation of phosphoric acid to a pH of 5.8. The nutrient solutions were prepared in consecutive steps. In total, 750 L of nutrient solution (250 L per tank) for each treatment (control, AP*unt* and AP*dis*) was prepared, respectively (1 L stock solution per 100 L nutrient solution). Before the experiment started, a sample from each nutrient solution tank was taken and analysed by ICP-OES and CFA (see 8.3.2.1) to determine the nutrient composition. The results are expressed in mg L^{-1}.

8.3.1.2 Fish water

The fish water used to prepare the nutrient solution for both aquaponic treatments was taken from the aquaculture unit of the DRAPS located at the Leibniz-Institute of Freshwater Ecology and Inland Fisheries (IGB) in Berlin, Germany, and is described in detail by Monsees et al. (2017b). The RAS unit was stocked with Nile tilapia (*Oreochromis niloticus*) at a mean stocking density of 90.6 kg m^{-3} in a total rearing volume of 5.1 m^3 resulting in a total fish biomass of 462.1 kg. The pH of the water was kept at around 7.2 and controlled by addition of calcium hydroxide, the temperature was 25 °C and the oxygen concentration remained always above 5.5 mg L^{-1}. Fish were fed with floating pellets (ALLER SANA FLOAT, Emsland Aller Aqua GmbH; Golßen, Germany) containing 37% crude protein, 10% crude fat, 38.5% nitrogen-free extractive, 6% crude ash, 3.5% crude fibre and 17.9 MJ kg^{-1} digestible energy. The fish were fed at a moderate feeding rate of 0.65% of the bodyweight per day. The study was carried out in compliance with the German legislation as authorized by the Regional Office for Health and Social Affairs Berlin (permit #: ZH 114).

8.3.2 Sampling of the nutrient solutions

The samples for the different analyses were taken on a weekly basis (pH, EC, colony forming units (CFU)) or three times during the experiment (TOC, DOC). Except for CFU (see 8.3.2.4) the first samples were taken at 21st March 2016 (3 days after planting) and marked as day or week 1, respectively. Samples for CFU were taken directly after disinfection (ad) (16th March 2016). Samples for TOC and DOC were taken at 21th March (week 0), in the 4th and at the last day of the experiment (week 7).

8.3.2.1 Nitrogen and nutrient analyses of process water and nutrient solutions

After collection all samples were frozen and stored in a freezer (-20 °C) until further analysis of nitrogen and mineral nutrients.

To determine concentrations of ammonium nitrogen (NH_4-N) and nitrate nitrogen (NO_3-N) in process water and nutrient solutions the samples were filtrated through 0.45 µm cellulose acetate membrane filters (GE Healthcare, United Kingdom) before the analysis using continuous flow analysis (CFA; San++, Fa. SKALAR; Breda, The Netherlands). The used method is described in detail by Suhl et al. (2016) and outlined in detail in the supporting material (S1-1).

The macronutrients (P, K, Ca, Mg, S, Na) and micronutrients (Fe, Mn, Mo, Zn, Cu, B) were analysed using inductively coupled plasma-optical emission spectrometry (ICP-OES, iCAP 6300Duo MFC, Fa. Thermo; Waltham, USA). After filtration, around 15 mL were used to flush the ICP-OES, and 45 mL were used for analysis. The measurement conditions (reference solution for calibration curves and wavelengths) for analysis of macronutrients are exactly described by Suhl et al. (2016) and Dannehl et al. (2015b) and are included in the supporting material (S1-2). Calibration curves for ICP-OES were made using micronutrient reference solutions with a concentration of 0–5 mg L^{-1} for Fe, Zn, Mn, Mo, Cu, and B.

8.3.2.2 pH and electrical conductivity

The pH and EC of the nutrient solutions were measured weekly (HI9811-5, Hanna Instruments Inc.; Rhode Island, USA) and occurring changes of the pH values were adjusted using sodium hydroxide (NaOH).

8.3.2.3 Total and dissolved organic carbon

For the analysis of total organic carbon (TOC) and dissolved organic carbon (DOC), two samples per nutrient solution were taken from each nutrient solution tank. For DOC analysis the samples were filtered through 0.45 µm cellulose acetate membrane filters (GE Healthcare, United Kingdom) being prepared by flushing with 20 mL distilled water and, subsequently with around 5 mL of the respective sample prior to sampling. The filtered DOC samples and the unfiltered samples for TOC analysis were subsequently acidified with 100 µL 2 M hydrochloric acid (HCl) and analysed using a TOC analyser following the NPOC-method according to the user's manual (TOC LCPNF Analyzer, Shimadzu; Kyoto, Japan).

8.3.2.4 Colony-forming units

For CFU analysis a dilution series (10^0, 10^{-1}, 10^{-2}) was prepared using casein soybean flour peptone bullion (CASO Bullion, Carl Roth GmbH & Co. KG; Karlsruhe, Germany). Each dilution was plated on 100 mm petri dishes filled with CASO-agar (Carl Roth GmbH & Co. KG; Karlsruhe, Germany) and incubated for 48 h at room temperature (20 °C). After incubation colonies were counted and CFU per mL was calculated. The whole work for CFU analysis was carried out in a microbiological safety cabinet (HeraSafe, Heraeus; Hanau, Germany).

8.3.3 Assessment of plant growth

At 10th May 2016, the lettuce heads were harvested by cutting the aboveground organ directly above the rock wool cube. Immediately after harvesting, each lettuce head was weighed to determine the mean fresh weight (FW) in gram. Afterwards, the leaves of nine lettuce heads per treatment (3 per individual gully, Figure 20) were cut and counted. The results were expressed as number of leaves per plant. Subsequently, the counted leaves were used to measure the leaf area per plant in dm^2 using an area meter (LI-3100 Area Meter; LICOR; Lincoln, NE, USA). The entire plant material was dried in a ventilated oven (Heraeus; Hanau, Germany) at 60 °C until constant weight to determine the dry weight in gram, which was used to calculate the dry matter (DM) %.

8.3.4 Sampling and preparation for chemical analysis of lettuce

For determination of nitrogen, carbon, mineral elements, nitrate, as well as the concentration of poly-phenolic compounds, four replicates per treatment were used. Each replicate included a mix of three lettuce heads, which were randomly sampled. Afterwards, the 12 lettuce heads per treatment were quartered. Three quarters of different lettuce heads were pooled as one sample so that all mixed samples contained different lettuce heads. The samples were shock frozen with liquid nitrogen and kept at -20 °C until they were freeze-dried for 48 h (Christ Alpha 1–4, Christ; Osterode, Germany) and ground to a fine powder. Each freeze-dried sample was divided into four sub-samples, one for elemental analysis (EA), one for inductively coupled plasma-optical emission spectrometry (ICP-OES), one for nitrate analysis, and one for high performance liquid chromatography (HPLC).

8.3.4.1 Nitrogen and carbon content as well as mineral elements in lettuce leaves

Nitrogen and carbon were determined according to DIN-ISO-10694 (1995) and DIN-ISO-13878 (1998) using an elemental analyser (vario MAX, Elementar Analysensysteme GmbH; Hanau, Germany). 0.3 g of the freeze dried samples was transferred into individual crucibles and catalytically combusted at 900 °C with pure oxygen. To separate nitrogen and carbon, the combusted sample and a carrier gas (helium) passed through specific adsorption columns at a specific temperature of 830 °C.

Afterwards nitrogen and carbon content were analysed by a thermal conductivity detector and determined according to the different thermal conductivity of nitrogen and carbon. The analysis was carried out twice and the results were calculated by the standard reference material glutamic acid and expressed at % DM.

For analysing mineral elements in lettuce leaves, 0.498 to 0.502 gram of a freeze dried sample were weighted into special, de-ionised vessels and subsequently analysed by ICP-OES. The sample preparation and digestion in the microwave (MARS Xpress, CEM; North Carolina, USA) as well as the analysis by ICP-OES are described in detail by Dannehl et al. (2012) and in supporting information (S1-3). The backup ensued by double determination and the results were expressed as mg 100 g^{-1} FW.

8.3.4.2 Nitrate content in lettuce leaves

Nitrate was analysed using a hand held reflectometer method (RQflex 10 plus, Merck KGaA; Darmstadt, Germany). For analysis 0.5 g of the powdered lettuce leaves were weighted into 50 mL volumetric flasks, filled up using distilled water and shaken for 20 min. Afterwards a barcode test strip was dipped into the sample solution for two seconds and the remaining solution was shaken off. During the reaction time of 60 seconds nitrate is reduced to nitrite. Nitrite reacts with an aromatic amine in the presence of an acid buffer to a diazonium salt. The diazonium salt in turn reacts with N-(1-Naphthyl)ethylenediamine to a red-violet azo dye. The quantity of the dye is measured by a reflectometer. The nitrate content was calculated and expressed in g kg FW^{-1}.

8.3.4.3 Analysis of phenolic compounds

For analysis of phenolic acids and flavonoids in lettuce leaves, 20 mg of the fine powder was extracted for 15 min in an ice water filled ultrasonic bath (Sonorex Super AG 102H, Bandelin electronic GmbH & Co. KG; Berlin, Germany) in 300 µL 70% methanol (pH 4.0, acetic acid). Afterwards, samples were centrifuged for 5 min at 4 °C at 10000 rpm (Multifuge X1R, Thermo Fisher Scientific; Waltham MA, USA) and the supernatant was collected in a 2 mLEppendorf tube. The pellet was extracted two times with 300 µL methanol (70%) for 10 min, respectively. The supernatants were concentrated in a vacuum concentrator evaporator (Savent SVC3000, Thermo Fisher Scientific; Waltham MA, USA) for around 2.5 h at 6 to 10 mbar. The resulting pellet was dissolved with 50% methanol and filled up to 1 mL. After shaking vigorously, 900 µL of the extract were filtered through a SpinX-Filter (Corning® Costar® Spin-X®) upon centrifugation for 5 min at room temperature and 3000 rpm. Finally, the extract was transferred into 2 mL HPLC-Vials and stored at -20 °C until analysis. 4-Meth-oxycinnamic acid (1 mM) and Apigenin-7-glucosid (0.5 mM) was used as internal standard for phenolic acids and flavonoids, respectively. For qualitative and quantitative analysis of the extracts the HPLC-system Ultimate 3000 (Thermo Fisher Scientific; Waltham MA, USA) was used as described by Förster et al. (2015) and is outlined in supporting material (S1-4). Phenolic acids and flavonoids were detected at a wavelength of 290 nm and 370 nm. The identification of phenolic acids and flavonoids was carried out according to the molecular mass (MS measurement performed on selected samples), UV spectrum, retention time and standards. The results were expressed as mg 100 g^{-1} FM.

8.3.5 *Calculation of fertilizer savings and greenhouse gas reduction*

To calculate the fertilizer saving potential, the amounts of single fertilizer (macronutrients) was calculated on a basis of g per L nutrient solution. The actual reduction was calculated referring to the control as 100%.
The reduction of greenhouse gases (expressed as CO_2-equivalents ($CO_{2\text{-}eq}$)) was calculated using the experimentally determined savings in inorganic fertilizers (aquaponic treatments) compared to the control (hydroponic treatment). The CO_2-equivalents were thereby calculated using the open source database ProBas of the Federal Environment Agency of Germany (UBA, 2015a). The exact data used for calculation and an example how the CO_2-equivalents were calculated for the respective fertilizers are listed in supporting material (S1-5).

8.3.6 *Statistical analysis*

For statistical analysis of all evaluated parameters the SPSS package version 19.0 was used. As such, to analyse the differences between the nutrient solutions (nutrient composition, pH, EC, TOC and DOC) as well as the plant parameters (FW, No. of leaves, leaf area, DM, nutrient and nitrate concentration, and phenolic compound) univariate analysis of variance (ANOVA) was used after normal distribution was confirmed using Shapiro-Wilk test. For post hoc testing Tukey-HSD (homogeneous variances) or Dunnett-T3 (heterogeneous variances) test was used. Not normal distributed values were compared using Kruskal-Wallis test and Dunn-Bonferroni test. CFU data were evaluated using Kruskal-Wallis test for the initial measurement after disinfection (ad) followed by a Friedman test for the analysis of the consecutive experimental weeks (week 0–7). Statistical analyses were carried out on a significance level of $p < 0.05$, where significant differences are indicated by different superscript small letters. When no superscript letters were inserted, no significant differences were detected. All data presented as mean values ± standard deviations of the respective samples.

8.4 Results

8.4.1 Plant yield and plant growth

The use of fish water, untreated or disinfected, resulted in similar yields as under conventional hydroponic conditions (Table 18). The lettuce heads reached a final fresh weight of 323.3 g (AP*unt*), 332.4 g (AP*dis*) and 325.9 g (C) after seven weeks of growth. Until harvest, the lettuce heads grown under control conditions formed 212.0 leaves. Lettuce heads in AP*dis* and AP*unt* formed 2 and 5.5 more leaves than the control, respectively. The formed leaf area was similar in all treatments and ranged between 66.7 dm^2 (AP*unt*) to 67.5 dm^2 (AP*dis*). The dry matter of lettuce grown under aquaponic conditions was 4.7%, 0.2% less than the control, though not significant.

Table 18: Growth parameters of butterhead lettuce grown in nutrient solution formulated with fresh water (control: 50:50, v/v fresh and rain water), fish water (untreated, AP*unt*) and disinfected fish water (AP*dis*). The fresh weight (FW), number of leaves (LN), leaf area (LA), and dry matter (DM) were compared and tested using univariate ANOVA. The data represent mean values of ten (FW) or three (LN, LA, DM) lettuce heads per repetition of each treatment.

Variant	Fresh weight (g)	Number of leaves	Leaf area (dm^2 $plant^{-1}$)	Dry matter (%)
Control	325.9 ± 53.6	212.0 ± 15.5	66.9 ± 5.7	4.9 ± 0.5
AP*unt*	323.3 ± 44.5	214.1 ± 42.3	66.7 ± 6.8	4.7 ± 0.3
APdis	332.4 ± 55.5	217.5 ± 29.4	67.5 ± 7.2	4.7 ± 0.2

No significant differences ($p < 0.05$) were detected by post hoc Tukey HSD test (FW) or Dunnett-T3 test (DM) as well as Kruskal-Wallis test between the different treatments.

8.4.2 Fertilizer and water savings

As evaluation criteria for improved environmental balance and to demonstrate the sustainability potential of the decoupled aquaponic system, the fertilizer savings for mixing the respective nutrient solutions were calculated (Table 19). The calculation was carried out based on single macronutrients only. As such, the total fertilizer supply could be reduced by 62.8% using fish water to mix the nutrient solution for lettuce. CAN fertilizer was identified as the main fertilizer which was saved (84.6%), followed by KNO_3 with a saving rate of 68.4%. Even the $MgSO_4$ supply was reduced by 36.4%. The addition of KH_2PO_4 was only reduced by 7.4%, corresponding to 2 g per L stock solution. In consequence of the reduced fertilizer demand in the AP treatments, the calculated amount of greenhouse gas emissions (expressed as $CO_{2\text{-eq}}$) by fertilizer manufacturing (incl. pre-process) could be reduced by 72.6% in comparison to the control. Additionally, in this setup, both aquaponic nutrient solutions were exclusively prepared with fish water, thus no additional freshwater was required (Table 19).

Table 19: Resource utilization (minerals, water) of butterhead lettuce grown in nutrient solution formulated with fresh water (control: 50:50, v/v fresh and rain water), fish water (AP*unt*) and disinfected fish water (AP*dis*).

Element	Unit[c]	Control[b]	AP*unt*/ AP*dis*[d]	Reduction[e] (%)	Control $CO_{2\text{-eq}}$ ($g\ L^{-1}$)[f]	AP*unt* / AP*dis*[d] $CO_{2\text{-eq}}$ ($g\ L^{-1}$)[f]
Fresh water[a]	**L**	375	0	100		
CAN	**$g\ L^{-1}$**	117	18	84.6	141.4	21.8
KNO_3	**$g\ L^{-1}$**	19	6	68.4	27.4	8.7
$MgSO_4$	**$g\ L^{-1}$**	44	28	36.4	13.0	8.3
KH_2PO_4	**$g\ L^{-1}$**	27	25	7.4	16.9	15.6
Total fertilizer supply/ reduction & $CO_{2\text{-eq}}$ emissions	**$g\ L^{-1}$, %**	**207**	**77**	**62.8**	**198.7**	**54.4**

[a]Total fresh water supply to all treatments; fresh water (50:50, v/v fresh and rain water).
[b]Control – fresh water prepared with mineral fertilizer; AP*unt* and AP*dis* – nutrient solution prepared with fish water and mineral fertilizer. AP*unt* was untreated and AP*dis* was disinfected nutrient solution.
[c]For fertilizer: g fertilizer use per L stock solution; 1 L stock solution is used to prepare 100 L nutrient solution.
[d]Same values for AP*unt* and AP*dis.*
[e]Control is 100%.
[f]CO_2-equivalent emission per L stock solution.

8.4.3 Nutrient composition of the used nutrient solution

The target nutrient concentrations for lettuce as recommended by Hochmuth (2001) (see 8.3.1.1) were established here for N, K, Mg, and most micronutrients. Still, Ca content was slightly higher, and Fe concentration was lower than recommended. At the beginning of the experiment, the analysis of the nutrient solutions revealed comparable equal mineral N (N_{min} = NO_3-N + NH_4-N) concentrations (Table 20). Nevertheless, the NH_4-N content was significantly higher in the control compared to both aquaponics treatments, whereas the NO_3-N contents were comparable. Except K and the already mentioned N forms, nutrient solutions revealed significant differences for all nutrients determined. However, significant differences below 10% are likely irrelevant and will not be presented in detail here. Most obvious, P was almost 2-fold increased in the control, compared to both aquaponic treatments. Also, Ca was substantially lower in the control (287 mg L^{-1}) than in both aquaponic treatments (319 and 322 mg L^{-1}). Further, Cu, Mo, Mn, Mg and Zn differed between groups. Here, Mg, S, Mo, and Zn were significantly lower in the control than in both aquaponic treatments.

Additionally, Na, Fe, Mn, Cu, and B were significant reduced in the control compared to at least one aquaponic treatment. Especially Fe and S were significantly decreased by 15-fold and 3.3-fold in the control.

Table 20: Nitrogen and mineral concentrations of the nutrient solution in mg L^{-1} at the beginning of the experiment used for control and both aquaponic treatments. The data represent the mean concentrations of nutrients in the three different nutrient solutions ± standard deviation measured after the nutrient solutions were mixed and completely prepared before the experiment started. The control were prepared with fresh and rain water (50:50, v/v), AP*unt* and AP*dis* were prepared with fish water. While AP*unt* was untreated, AP*dis* was disinfected. The data was compared and tested using univariate ANOVA and post hoc tests Tukey-HSD test or using Kruskal-Wallis (Cu) and Dunn-Bonferroni (Fe, Mo, B) test. Different superscript small letters indicate significant differences between the treatments ($p < 0.05$).

Element	Control	AP*unt*	AP*dis*
NO_3-N	191.4 ± 8.2^{a}	197.1 ± 10.5^{a}	191.2 ± 22.9^{a}
NH_4-N	14.0 ± 0.9^{c}	2.2 ± 0.01^{a}	2.3 ± 0.01^{b}
N_{min}[1]	205.4 ± 9.1^{a}	199.4 ± 10.5^{a}	193.5 ± 23.0^{a}
P	144.8 ± 0.5^{b}	73.4 ± 0.9^{a}	74.5 ± 0.4^{a}
K	148.0 ± 1.5^{a}	145.8 ± 1.5^{a}	146.4 ± 1.9^{a}
Ca	286.7 ± 3.7^{a}	318.7 ± 3.6^{b}	322.3 ± 1.0^{b}
Mg	52.0 ± 0.5^{a}	56.0 ± 0.3^{b}	57.4 ± 0.5^{c}
S	83.8 ± 1.6^{a}	137.8 ± 1.3^{b}	142.4 ± 0.9^{c}
Na	55.2 ± 0.3^{a}	55.6 ± 0.3^{a}	57.0 ± 0.5^{b}
Fe	0.001 ± 0.00^{a}	0.01 ± 0.00^{ab}	0.02 ± 0.00^{b}
Mn	0.44 ± 0.00^{a}	0.45 ± 0.01^{b}	0.44 ± 0.00^{a}
Mo	0.02 ± 0.00^{a}	0.03 ± 0.00^{b}	0.03 ± 0.00^{b}
Zn	0.03 ± 0.01^{a}	0.05 ± 0.01^{b}	0.06 ± 0.01^{b}
Cu	0.04 ± 0.04^{ab}	0.03 ± 0.00^{a}	0.06 ± 0.00^{b}
B	0.24 ± 0.00^{a}	0.28 ± 0.00^{ab}	0.29 ± 0.00^{b}

[1]N_{min} is the sum of NO_3-N and NH_4-N.
[a,b,c]significant differences are indicated by different superscript small letters ($p < 0.05$).

8.4.4 pH value and electrical conductivity of the nutrient solution

The weekly control of pH and electrical conductivity (EC) showed that the pH in the nutrient solution of the control started to drop down to 5.37 and 5.13 in the 3rd and 4th week, respectively (Figure 21). Still, differences to both aquaponic treatments were not significant. In the 5th and 6th week the pH levelled again at around 6 before it decreased again in week 7. However, the pH of the nutrient solution based on fish water was relative uniform and constant and varied only between 5.5 and 5.8. In contrast, the EC level of all nutrient solutions varied stronger than the pH, but the development of the EC over time was similar in all treatments (Figure 21). As such, a more or less continuous increasing EC value from the beginning of the experiment (2.2 dS m^{-1}) to the 7th week (around 3.0 dS m^{-1}) was observed. Nevertheless, the differences between the different treatments were significant from the 1st to the 5th week. The EC in AP*dis* was always significantly higher than in the control, except in week 1[13]. The EC value of both aquaponic treatments did not differ over the experimental period. With an increasing standard deviation, the EC values did not differ significantly from each other in the last two weeks of the experiment.

[13] Taken from original publication; correct would be 1, 6, and 7.

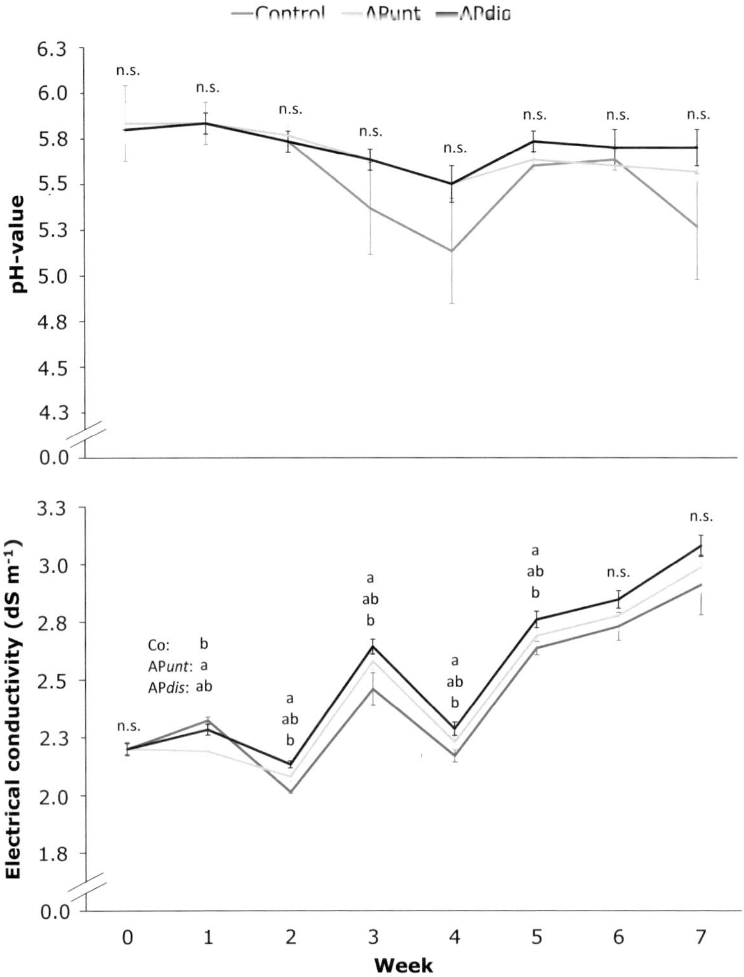

Figure 21: pH and electrical conductivity (EC) of the nutrient solutions formulated with fresh water (control: 50:50, v/v fresh and rain water), fish water (AP*unt*) and disinfected fish water (AP*dis*) over the experimental period of 8 weeks. The data represent mean values of three replicates ± standard deviation, measured once a week. The mean values were compared using univariate ANOVA and tested using post hoc tests Tukey-HSD (EC in week 3, 6, and 7) or compared using Kruskal-Wallis test (pH all and EC in week 0) and following post hoc Dunn-Bonferroni ($p < 0.05$). Different small letters indicate significant differences between the three different nutrient solutions at the respective time point and are listed from top to bottom: control, AP*unt,* and AP*dis*.

8.4.5 Total and dissolved organic carbon

The measurement of the TOC and DOC showed significant differences between the different nutrient solutions (Table 21). With the use of fish water, the aquaponic treatments contained concentrations at the beginning of the experiment for TOC 3.4-fold (AP*dis*) and 3.5-fold (AP*unt*) and for DOC 3.8-fold (AP*dis*) and 4.1-fold (AP*unt*), compared to the control. In the following weeks the differences decreased down to 2.9-fold (AP*unt*) and 3.0-fold (AP*dis*) for TOC and 3.2-fold (AP*unt*) and 3.4-fold (AP*dis*) for DOC, respectively. The disinfected nutrient solution (AP*dis*) had a significantly reduced DOC content compared to the untreated aquaponics (AP*unt*) at the beginning of the experiment. However, in the following weeks, the values dropped and were not significantly different any more.

Table 21: Total (TOC) and dissolved organic carbon (DOC) in the different nutrient solutions. The values represent mean values of three replicates ± standard deviation. Different superscript small letters indicate significant differences between the three treatments within one measuring week. The mean values were compared using univariate ANOVA and differences were tested using Dunnett-T3 test ($p < 0.05$). The control was prepared with fresh and rain water (50:50, v/v), AP*unt* and AP*dis* were prepared with fish water. While AP*unt* was untreated, AP*dis* was disinfected.

Week	Parameter	Control	AP*unt* (mg L^{-1})	AP*dis*
0	**TOC**	7.4 ± 0.85^{a}	26.1 ± 0.44^{b}	25.5 ± 0.59^{b}
	DOC	5.9 ± 0.08^{a}	24.3 ± 0.15^{c}	22.6 ± 0.15^{b}
4	**TOC**	8.0 ± 0.93^{a}	23.1 ± 0.80^{b}	24.2 ± 1.72^{b}
	DOC	7.0 ± 0.87^{a}	22.1 ± 0.94^{b}	22.7 ± 1.17^{b}
7	**TOC**	8.6 ± 0.97^{a}	27.6 ± 2.25^{b}	28.4 ± 0.54^{b}
	DOC	8.1 ± 0.89^{a}	26.8 ± 2.36^{b}	27.9 ± 0.55^{b}

^{a,b,c}significant differences are indicated by different superscript small letters ($p < 0.05$).

8.4.6 Colony-forming units content in the nutrient solutions

No CFU were detected in AP*dis* after disinfection (Figure 22). The control started with an average CFU content of 177 colonies per mL^{-1} and contained significant more CFU than AP*dis* but significant less than the untreated fish water based nutrient solution (5097 CFU mL^{-1}). From the day of disinfection until the first measurement on 21st March 2016 (week 0, 5 days after disinfection, 3 days after planting; see 8.3.1), the CFU in AP*unt* and AP*dis* increased rapidly. In the following two weeks the CFU in both aquaponic treatments dropped down, but varied strongly between the replicates, as revealed by a high standard deviation. In the 3rd and 4th week the nutrient solutions of all treatments dropped to a relatively stable level of around 662 (AP*unt*) to 1518 (AP*dis*) CFU mL^{-1}. From the 5th week until the end of the experiment the CFU decreased further and ranged between 502 (AP*unt*) and 949 (AP*dis*) in all treatments. The Friedman test did not indicate significant differences between the treatments.

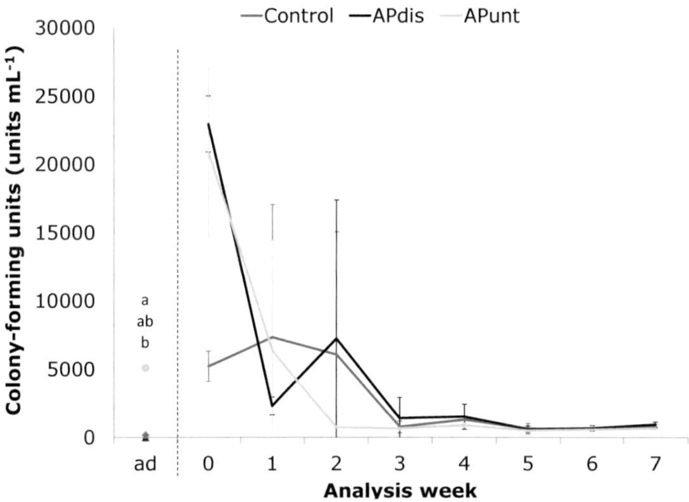

Figure 22: Content of colony-forming units (CFU) of the nutrient solution formulated with fresh water (control: 50:50, v/v fresh and rain water), fish water (AP*unt*) and disinfected fish water (AP*dis*) over the experimental period of 7 weeks. The data represent mean values of three repetitions per treatment ± standard deviation. Mean values were evaluated using Kruskal-Wallis test for the time after disinfection (ad) and by the Friedman test for the analysis of the consecutive experimental weeks (week 0–7) at a level of $p < 0.05$. Significant differences are indicated by different small letters and listed from top to bottom in the following range: AP*unt,* control, and AP*dis*.

8.4.7 Nitrate, nitrogen, carbon, and mineral content in lettuce leaves

Lettuce leaves from the three treatment groups revealed significantly different contents of Ca, Mg, S, and Na (Table 22) on a dry mater basis as analysed by ICP-OES analysis of macro- and micro elements. Here, the Ca and Mg was significantly different in-between both aquaponics treatment but not to the control. Both were increased in the untreated aquaponics treatment (Ca = 9.8% and Mg = 10.7%). The content of S was significantly decreased (7.1%) in the conventional nutrient solution compared to AP*dis.* Also, Na was lower by 18.5% and thus significantly less in AP*unt* compared to the control. The amount of Fe was not significantly different between the treatments.

Table 22: Effect of fresh water (control, prepared using fresh and rain water (50:50, v/v), untreated fish water (AP*unt*) and disinfected fish water (AP*dis*) based nutrient solution on nitrate and mineral content in butterhead lettuce. The data represent mean values on dry weight matter (DM) or fresh weight basis (FW, only for NO_3) ± standard deviation of four mix samples taken randomly from respective three lettuce heads. Mean values were compared using univariate ANOVA and following tested using post hoc Tukey-HSD test or compared by Kruskal-Wallis test (P) and tested using Dunn-Bonferroni test (Mg, Na) ($p < 0.05$). Significant differences are indicated by different superscript small letters.

Element	**Unit**	**Control**	**AP*unt***	**AP*dis***
NO_3	**mg kg^{-1} FW**	5896.3 ± 453.7^{a}	5795.3 ± 456.3^{a}	5502.1 ± 367.7^{a}
C	**%DM**	42.25 ± 0.44^{a}	42.48 ± 0.59^{a}	42.0 ± 0.34^{a}
N	**%DM**	4.12 ± 0.19^{a}	4.29 ± 0.12^{a}	4.27 ± 0.09^{a}
P	**%DM**	0.71 ± 0.03^{a}	0.73 ± 0.02^{a}	0.73 ± 0.03^{a}
K	**%DM**	6.33 ± 0.21^{a}	5.94 ± 0.16^{a}	6.29 ± 0.32^{a}
Ca	**%DM**	1.79 ± 0.19^{ab}	1.91 ± 0.06^{b}	1.74 ± 0.06^{a}
Mg	**%DM**	0.28 ± 0.03^{ab}	0.31 ± 0.00^{b}	0.28 ± 0.01^{a}
S	**%DM**	0.26 ± 0.01^{a}	0.27 ± 0.01^{ab}	0.28 ± 0.01^{b}
Na	**%DM**	0.27 ± 0.02^{b}	0.22 ± 0.01^{a}	0.21 ± 0.01^{ab}
Fe	**%DM**	0.59 ± 0.06^{a}	0.60 ± 0.10^{a}	0.57 ± 0.02^{a}

8.4.8 Phenolic acids and flavonoids in lettuce leaves

The analysis of the different phenolic acids and flavonoid glycosides showed that the investigated green butterhead lettuce contained mainly dicaffeolyltartaric (between 2.45 and 3.44 mg $100g^{-1}$ FW) (Table 23). The other phenolic acids did not follow a consistent range. The lowest concentrations were detected for coumaroylquinic acid ranging between 0.028 mg $100g^{-1}$ FW in control lettuce and 0.038 mg $100g^{-1}$ FW in AP*unt*. The content of the different phenolic acids, as well as the analysed flavonoid quercetin-3-O-(6´´-O-malonyl) did not differ significantly.

Table 23: Concentrations of phenolic acids and flavonoid glycosides in mg per 100 g fresh weight in butterhead lettuce caused by different nutrient solutions. The nutrient solutions for control were prepared with fresh and rain water (50:50, v/v) and for AP*unt* and AP*dis* with fish water. The latter were untreated for AP*unt* and disinfected for AP*dis*. The data represent mean values of four lettuce heads samples, mixed in turn of three different lettuce heads ± standard deviation. No significant differences ($p < 0.05$) were analysed using univariate ANOVA and Tukey-HSD test (caffeoyltartaric acid, caffeoylquinic acid, and caffeoylmalic acid) or using Kruskal-Wallis test.

Metabolite	Control	AP*unt* (mg 100 g^{-1} FW)	AP*dis*
Phenolic acids			
caffeoyltartaric acid (CTA)	0.97 ± 0.23[a]	1.22 ± 0.34[a]	1.29 ± 0.29[a]
caffeoylquinic acid (chlorogenic acid)	0.93 ± 0.15[a]	1.08 ± 0.24[a]	1.12 ± 0.24[a]
coumaroylquinic acid	0.028 ± 0.005[a]	0.033 ± 0.013[a]	0.038 ± 0.010[a]
caffeoylmalic acid (CMA)	1.07 ± 0.22[a]	1.13 ± 0.36[a]	1.34 ± 0.36[a]
dicaffeoyltartaric acid (chicoric acid)	2.45 ± 0.58[a]	3.26 ± 0.74[a]	3.44 ± 0.75[a]
Flavonoid			
quercetin-3-O-(6´´malonyl)glucoside (Q3MG)	0.25 ± 0.03[a]	0.33 ± 0.07[a]	0.36 ± 0.05[a]

8.5 Discussion

8.5.1 Plant yield and plant growth

As one of the most important and obvious results of the present study, the fresh weight (FW) of butterhead lettuce did not differ significantly between both aquaponics treatments and the conventional hydroponic control (Table 18). Consequently, a similar number of leaves, leaf area and dry matter content of all lettuce heads were observed. These results are not surprising as nutrient solutions had comparable nutrient concentrations. As a logical consequence, the growth of plants should be comparable. As hypothesized by Delaide et al. (2016), RAS derived carbon sources or microorganisms might have beneficial effects on plant growth cannot be supported in the present study. Nevertheless, it was found that decoupled aquaponics are as effective as conventional hydroponics but using around 63% less mineral fertilizer and 100% less fresh water for lettuce production

(Table 19). Similar was found by Suhl et al. (2016) and Suhl et al. (2018b) who documented no significant differences in tomato yield between conventional hydroponics and supplemented aquaponics. In contrast, Delaide et al. (2016) reported that the supplementation of fish water with mineral fertilizer increased the FW of lettuce by nearly 40%. Delaide et al. (2016) concluded that substances with plant-promoting effects, like humic-like and protein-like dissolved organic matter (DOM) or rhizobacteria and/or fungi may have enhanced plant growth. According to this hypothesis, the present study should have resulted in differences between conventional and aquaponic treatments. As such, the growth-promoting effects of DOC would have increased the growth in untreated and disinfected aquaponics (AP*unt* und AP*dis*) compared to the control. Also, growth stimulation due to beneficial bacteria or fungi would have resulted in differences between AP*unt* and the control and potentially also between both aquaponic treatments. Adani et al. (1998) found beneficial effects of humic acids on plant growth in hydroponics. As reported, humic acids added to hydroponics may have different effects on plant growth, e.g. an positive effect on shoot:root ratio (Adani et al., 1998). The DOC determined in the waste water of RAS includes humic substances (Leonard et al., 2002). Humic substances are highly complex substances containing humid acids as extractable fractions (MacCarthy, 2001) and the accumulation of humic substances in RAS depends on the feed input (Hambly et al., 2015; Leonard et al., 2002) and/or the fresh water exchange rate (Leonard et al., 2002). Therefore, the composition of humic substances may vary and might be different from RAS to RAS, and study to study. As such, the question on how the DOC, the produced fish species and/or the composition of the fish feed might improve the above ground biomass of lettuce can result in divergent findings. Additionally, the current approach used a static nutrient supply as routinely used in lettuce production. For tomato or cucumber culture, a discontinuous refilling of the nutrient reservoir is necessary due to the extended culture period of these crops. This could eventually also result in different growth patterns such as improved growth in the aquaponic treatment compared the conventional hydroponic production since the exposure time of RAS derived microorganisms and organic carbon sources are prolonged.

8.5.2 Fertilizer savings and fertilizer use efficiency

Around 63% less mineral fertilizer was used in the aquaponic treatments compared to the hydroponic control (Table 19). Due to the nutrient composition of the fish water, especially CAN fertilizer addition was reduced (-84.6%) by decoupled aquaponic treatments. In contrast, the lowest reduction was realised with the phosphorus fertilizer (KH_2PO_4), which resulted in a reduction of 7.4%. Fertilizer production is an energy demanding process. As Jenssen and Kongshaug (2003) reported, approximately 1.2% of the world´s energy consumption is assigned to fertilizer production which is responsible for 1.2% of the world´s greenhouse gas (GHG) emissions. GHG emissions originate e.g. from the manufacturing processes of nitrogen-containing fertilizer or from fossil fuel utilization for the production and the transport of raw material (Wood & Cowie, 2004). As a consequence, by reducing fertilizer use and decreasing fertilizer production, GHG emissions could be reduced substantially. Arguably, this is only the case if used as add-on for pre-existing or newly planned RAS, which primarily serve the purpose of satisfying an increasing demand for animal protein. Aquaponic systems could, due to the increased circularity and the recycling of nutrients, substantially reduce the demand for inorganic fertilizer, and thus lower GHG emissions compared to conventional hydroponic systems. However, considering a theoretical global dietary shift in the future towards plant based diets, one could argue that the energy requirements for e.g. building and operating a RAS, feed production, etc. will release the same or even higher GHG emissions compared to the production of inorganic fertilizer. Here, a RAS would primarily serve the purpose of nitrogen production.

Due to the large variety of fertilizers the direct calculation of GHG emissions from fertilizer production is difficult (Wood & Cowie, 2004). According to the calculation (see supporting material S1-5) using ProBas database (UBA, 2015b), the amount of $CO_{2\text{-eq}}$ produced by 1 L stock solution was reduced by 72.6% using fish water to prepare the nutrient solution (Table 19). Considering a lettuce yield of 41 kg m^{-2} $year^{-1}$ and a water use of 20 L kg^{-1} $year^{-1}$ (Barbosa et al., 2015) the total nutrient solution demand per year would be 8 200 m^3 per ha (i.e. 82 m^3 fertilizer stock solution, compare Table 19). Extrapolated to an annual lettuce production in a one hectare greenhouse, the emitted GHG (manufacturing) would amount 16.3 t $CO_{2\text{-eq}}$ for the conventional hydroponic production (control). In contrast, using nutrient enriched fish water could reduce this amount to only 4.5 t CO_{2-eq} (-72%) as shown in the present study.

8.5.3 Nutrient management in decoupled aquaponics

The process water for mixing the nutrient solutions for the aquaponics treatments was used directly from the RAS without prior dilution to reflect realistic, practical conditions. Here, it was possible to adjust most nutrients to nearly similar concentrations (Table 20). Nevertheless, some nutrients differed due to major differences in the aquaponic RAS and the tap/rain-water mixture as well as the standard fertilizer used here.

Similar to the finding of Delaide et al. (2016), several elements were decreased, most prominent NH_4^+ (but not N_{min}), P (2-fold) or increased Ca, S, Fe, Zn in the aquaponics compared to the control. Nitrogen is a decisive nutrient for plant growth (Hawkesford et al., 2012). Therefore, the focus for mixing the nutrient solution in the present study was to align N in the different treatments, which was only possible with regard to the total N. The nearly 2-fold higher P content in the control can be explained by the necessity of a much higher addition of phosphoric acid due to the higher alkalinity of the tap water to drop the pH to 5.8. In general, RAS process water often has a decreased alkalinity due to nitrification processes in the biofilter (release of protons) and due to an increased input of CO_2 (respiratory end-product of fish and microorganisms). Nitric acid which is commonly used to lower the pH (Snyder, 2016) would have increased the nitrogen content, which was undesirable. As a consequence of the levelling of N, the Ca content in all treatments was higher than recommended by Hochmuth (2001) because it was added in combination with nitrate as calcium-ammonium-nitrate (CAN) fertilizer. The S content was 1.6-fold higher in both aquaponic treatments as a result of Mg equilibration by $MgSO_4$. The significant differences between both aquaponic treatments were minor, though partially significant. Higher values upon disinfection (NH_4-N, Mg, S, Na, and Cu) may be explained by the increased evaporation of water.

The interactions of soluble nutrients in the root zone itself can have effects on nutrient uptake by plants. Such interactions between nutrients are diversified (Fageria, 2001): While an increasing amount of specific nutrients, for example reported for S (Datko et al., 1978) and Zn (Coolong & Randle, 2003), results in an increased accumulation of these nutrients in plants, other nutrients can have synergetic and antagonistic effects

(Fageria, 2001). However, the influence of divergent nutrient solutions, without being more specific, resulted in a different nutrient content of the plants, not only between the control and at least one aquaponic treatment, but also between both aquaponic treatments (Table 22). The mineral content in all treatment groups showed no obvious deficiencies compared to the ranges reported by Hochmuth et al. (2004). Thus, significant differences in the nutrient concentrations of the different treatment groups had obviously no negative effects on the plant growth.

Due to the highly complex interactions of ions and the differences in the nutrient solutions, slightly different concentrations in the crop may occur but did ultimately not affect the plant growth here. It is likely that the nutrient profile in decoupled aquaponics will always differ slightly from a recommended nutrient solution for hydroponic plant production. As shown in this study, disadvantages for aquaponic plant production are not expected.

8.5.4 Abiotic and microbiological features of the nutrient solution

The less stable pH in the nutrient solution of the control (Figure 21) may be a result of the higher ratio of NH_4-N (Table 20). It is known, that the uptake of NH_4-N is related to an excretion of protons by the plant roots (Hawkesford et al., 2012), whereby the nutrient solution becomes acidic. The addition of NaOH was effective in increasing the pH to reach target values without adding further nutrients (NPK) except of Na. A slight but continuously increasing EC value in the nutrient solution during the eight weeks of experiment, as detected in all treatments, is commonly reported as a result of accumulation of Na and Cl and macronutrients like Ca, Mg, K (Savvas et al., 2007), and sulphate (Zekki et al., 1996).

As expected, the TOC and DOC content were significantly increased by using fish water to mix the nutrient solution. DOC in the control was in a range with already reported values by Waechter-Kristensen et al. (1999). However, as already mentioned, it seems that the increased TOC/DOC content in the aquaponics treatments did not have a positive effect on plant growth here.

The disinfection of the fish water based nutrient solution was considered to be successful because no CFU were detected in AP*dis* after disinfection (Figure 22). Nevertheless, CFU although commonly used, is just an estimate of the most probable number of bacteria as several aquatic microbes are not cultivable in standard media (Amann et al., 1995).

Generally only 1% of the waterborne microorganisms in ground and drinking water are cultivable *in vitro* (Ultee et al., 2004). But a thermal treatment of > 90 °C for more than 15 min is an accepted procedure for disinfection (Ehret et al., 2001) and additionally the temperature of the nutrient solution was maintained > 70 °C for more than 45 min. Two days after successful disinfection lettuce plants were planted and the nutrient flow started. After the first five days after disinfection (on 21st March 2016) the CFU had increased rapidly to around 2.1 x 104 (AP*unt*) and 2.3 x 104 (AP*dis*) units mL^{-1} in both aquaponic treatments. Compared to the reported CFU in RAS, rearing marine see bass (*Dicentrarchus labrax*) (Leonard et al., 2000), the total content of CFU in fish water was relatively low in the present study. However, the microbial community and its density depends on fish species and varies from RAS to RAS (Sugita et al., 2005). As stated earlier, CFU is an estimate of the bacteria present in the nutrient solution; it is not reflecting the total abundance of microorganisms, since most of these organisms are not cultivable (Ehret et al., 2001). The reason for the high number of CFU might be the infection of the nutrient solution by spores or external microorganisms from the air and from the roots. The roots and its environment for instance, can be densely populated by microorganisms (Waechter-Kristensen et al., 1999) which are easily spread within the water body. Therefore, disinfection will soon allow the establishing of a new microbiota. The increase of the CFU within the first five days after planting was less strong in the control compared to both aquaponics treatments and resulted in a significant lower content at the first measurement, suggesting suboptimal conditions for bacteria (i.e. heterotrophic bacteria) compared to the aquaponic groups. Usually organic matter (DOC and TOC) is tightly correlated with bacterial growth. Most probably, lower DOC/TOC concentrations might have restricted growth in the control. Nevertheless, in all three treatments a classical growth pattern (lag-, log-, stationary- and death phase) was observed. Still the lag- and the stationary phase were not so pronounced in our experiments which could be explained by the weekly sampling intervals for CFU (Madigan et al., 2017). At the end of the experiment the CFU ranged between 502 (AP*unt*) and 949 (AP*dis*) in all treatments as it was similarly reported for sterilized and non-sterilized aquaponic lettuce production by Pantanella et al. (2015).

8.5.5 Effect on quality parameters of lettuce

8.5.5.1 Phenolic compounds in lettuce leaves

It is widely recognised that plant-derived phenolic compounds and their antioxidant activity have beneficial health effects to humans (Pandey & Rizvi, 2009). Thus, phenolic compounds represent a commonly accepted quality parameter for the evaluation of plants. Phenolic compounds are ubiquitous in plants. In humans, they are mainly ingested via fruits and beverages. Vegetables are an important polyphenol source in the human diet (Scalbert & Williamson, 2000). As such, the intake of antioxidant rich fruit and vegetables correlates inversely with the incidence of cancer (Boudet, 2007). Lettuce itself showed strong antioxidant and anti-inflammatory activities in *in vitro* studies (Mulabagal et al., 2010). However, the content of polyphenols is influenced by different abiotic factors such as e.g. light, temperature, water management, carbon dioxide concentration, and fertilisation (Dannehl & Josuttis, 2014). The composition of phenolic compounds determined in the present study were already described for green lettuce and within the reported levels by DuPont et al. (2000), Llorach et al. (2008), and Ribas-Agustí et al. (2011). Applying different nutrient solutions as in the present study did not result in significant differences. However, the nutrient composition in the soil can have an effect on the content of phenolic acids and flavonoids (Becker et al., 2015; Radi et al., 2003; Theunissen et al., 2010). An increasing N content, for instance, decreases the content of phenolic acids and flavonoids in different plant species (Becker et al., 2015; Radi et al., 2003), whereas an increasing K content results in increasing phenolic compounds (Radi et al., 2003). Both, N and K did not differ in the nutrient solution used for the different treatments in the present study. Here, the biggest differences were detected for P and S (Table 20). It was found by Radi et al. (2003) that P fertilizer did not have a strong effect on phenolic compounds in comparison to N and K. S, however, mainly effects S containing secondary metabolites like glucosinolates (Falk et al., 2007), which is generally found in *Brassica* species. Furthermore, it has been reported that high molecular humic substances can stimulate the phenylpropanoid metabolism resulting in an accumulation of specific phenolics in maize (*Zea mays*) (Schiavon et al., 2010). However, the use of different nutrient solutions with it's partly varying nutrient concentration (Table 20), different TOC/DOC content (Table 21), and a probably different bacterial composition, as used in the present study did not affect the analysed phenolic acids and flavonoids (Table 23).

8.5.5.2 Nitrate content in lettuce leaves

The nitrate concentration is an important quality parameter for lettuce. It can have negative effects to human health when it is converted to nitrite by microorganisms (Lundberg et al., 2004). In Germany a maximum level of 4000 (harvest from 01.10. until 31.03.) to 5000 mg NO_3 kg^{-1} FW (harvest from 01.04. until 30.09.) is accredited (VO (EU) Nr. 1258/2011, 2011). Thus, the lettuce in the present study contained more nitrate (5502.1 to 5896.3 mg NO_3 kg^{-1} FW) than accredited by VO (EU) Nr. 1258/2011 (2011) (Table 22). Nevertheless, concentrations up to 4500 and 5406 mg NO_3 kg^{-1} FW were also reported from other authors for fresh lettuce (Anjana & Iqbal, 2007; Merino et al., 2006). The nitrate content of lettuce grown in hydroponic can be reduced by a combination of nitrate and ammonium nutrition (Demsar & Osvald, 2003). As such, in contrast to the results in the present study it was expected that the higher ammonium content in the control resulted in a reduced nitrate content. One reason that the higher NH_4-N content had no effect on the nitrate content in the leaves might be that the ammonium concentration was too low to influence the nitrate content significantly. Demsar and Osvald (2003) , for instance, found a reduced nitrate content of 13% by feeding at a NO_3:NH_4 ratio of 6.6:1. In the present study, the content of ammonium was much lower (13.7:1) and equal to other control treatments as reported by Gunes et al. (1995). Another reason for similar nitrate content in the control and the aquaponic treatments might be assigned to the higher amount of DOC in aquaponics compensating the ammonium effect in the control. As such, humic acid can reduce the accumulation of nitrate of diverse vegetable crops (Nazaryuk et al., 2002). Furthermore, the nitrate content depends on growth period (Economakis & Koleilat, 1996) and nitrate availability for plants (Economakis & Koleilat, 1996; Konstantopoulou et al., 2010; McCall & Willumsen, 1998). Economakis and Koleilat (1996) found out, that the longer lettuce is feed with nitrate, the more it accumulates in the plants. An earlier harvest in the present study with a fresh weight below 300 g could have resulted in lower nitrate content. Additionally, the nitrogen concentration in the nutrient solution could have been adjusted. Probably, a lower nitrogen concentration

(Mattson & Peters, 2014) would have resulted in a lower nitrate concentration without affecting the yield (Economakis & Koleilat, 1996).

8.6 Conclusion

The benefits of nutrient and water recycling caused by a combined production of fish and plants in a decoupled aquaponic approach were clearly shown in this study. The recycling of fish water resulted in a reduction of fertilizer application of nearly 63% and a complete substitution of freshwater. No significant difference in growth of green open butterhead lettuce was observed, which is probably based on the same nutrient supply.

The content of DOC/TOC as well as microorganisms derived from RAS had no obvious positive/beneficial effects on the growth and phenolic compounds of lettuce.

The post adjustment of the nutrient profile using fish water for professional hydroponic application is more challenging as in conventional hydroponics (with rain water or tap water) but it was demonstrated that for most nutrients the set points were reasonable close to the recommended nutrient concentrations. With ongoing professionalization and standardisation of practises in decoupled aquaponic technology it is assumable that optimal nutrient profiles in decoupled aquaponic applications will be met with reasonable effort and comparable yields to conventional hydroponic production.

Additionally, butterhead lettuce (Salanova) has been proven to be a promising plant for potential intercropping as a further cash crop during the unproductive period for tomatoes in temperate regions, namely the winter.

8.7 Supporting material

8.7.1 S1-1 Nitrogen analysis of process waters and nutrient solutions

To determine the nitrogen content the samples were filtrated and analysed using continuous flow analysis (CFA; San++, Fa. SKALAR; Breda, The Netherlands). The used method is exactly describes by Suhl et al. (2016). As reference solution 0.584 mg L^{-1} ammonium standard solution (Merck KGaA; Darmstadt, Germany) and 0.565 mg L^{-1} nitrate standard solution (Merck KGaA; Darmstadt, Germany) were used. Dissolved ammonium nitrate was used to establish the standard series for CFA as following concentrations: 0.25 mg L^{-1}, 0.5 mg L^{-1}, 1.0 mg L^{-1}, 2.0 mg L^{-1} and 3.0 mg L^{-1}. Calcium chloride was used as rinse solution for the automatic sampler. The following working solutions for ammonium nitrogen (NH_4-N) were used:

buffer solution (66 g potassium sodium tatrate + 48 g sodium citrate trihydrate + 6 mL Brij (30%) in 2 L distilled (DI) water), sodium salicylate (12.5 g NaOH + 40 g sodium salicylate in 0.5 L DI water), sodium nitroprusside (0.5 g in 0.5 L DI water) and sodium dichloroisocyanurate (1.0 g in 0.5 L DI water). The measurement of NH_4-N was performed at a wavelength of 660 nm. For nitrate nitrogen (NO_3-N) analysis, a buffer solution with 3.405 g imidazole diluted in HCl and 3 mL Brij (30%) diluted in 1 L DI water were blended with a staining reagent with 50 mL H_3PO_4, 5 g sulfanilamide and 0.25 g naphthylethylendiamindihydrochloride diluted in 0.5 L DI water. The measurement was performed at a wavelength of 540 nm.

The different nitrogen forms were analysed in duplicate and expressed as mg L^{-1}. The nitrogen forms NH_4-N and NO_3-N were additional summarized to mineral N (N_{min}).

The respective microelements in the digestion solution were analysed at following wavelengths: Fe = 259.9 nm, Mn = 257.6 nm, Mo = 202.0 nm, Zn = 206.2 nm, Cu = 324.8 nm, and B = 249.8 nm.

The different nitrogen forms and the mineral elements were analysed in duplicate and expressed as mg L^{-1}. Both nitrogen forms NH_4-N and NO_3-N were summarized as mineral N (N_{min}).

8.7.2 S1-2 Nutrient analysis of process waters and nutrient solutions

The makro- (P, K, Ca, Mg, S, Na) and micronutrients (Fe, Mn, Mo, Zn, Cu, B) in the process waters and nutrient solutions were analysed using inductively coupled plasma-optical emission spectrometry ((ICP-OES), iCAP 6300Duo MFC, Fa. Thermo; Waltham, USA). After filtration, around 15 mL were firstly used to flush the ICP-OES, and 45 mL were following used for measurement. The following reference solutions were used to assemble the calibration curve for ICP-OES: blank 1.4 molar HNO_3; 0–100 mg L^{-1} for P, Mg and Na, 0–300 mg L^{-1} for K and Ca, , 0–50 mg S L^{-1}, and 0–5 mg L^{-1} for Fe, Zn, Mn, Mo, Cu, and B. The respective elements in the digestion solution were analysed at following wavelengths: P = 213.62 nm, K = 766.5 nm, Ca = 317.9 nm; Mg = 279.1 nm; S = 182.0 nm, Na = 589.6 nm, Fe = 259.9 nm, Mn = 257.6 nm, Mo = 202.0 nm , Zn = 206.2 nm, Cu = 324.8, and B = 249.8 nm.

The different mineral elements were analysed in duplicate and expressed as mg L^{-1}. The nitrogen forms NH_4-N and NO_3-N were additional summarized to mineral N (N_{min}).

8.7.3 S1-3 Mineral elements in lettuce leaves

For analysing mineral elements in lettuce leaves, the freeze dried samples were analysed by ICP-OES. In detail, for sample preparation, between 0.498 and 0.502 g of the sample material was weight into special, deionised vessels. The vessels were placed into a microwave (MARS Xpress, CEM; North Carolina, USA) after the addition of 5 mL nitric acid (HNO_3, 65%) and 3 mL hydrogen peroxide (H_2O_2, 30%). The digestion was carried according to following program: 1st step = 20 min to reach 200 °C, 2nd step = 5 min at 200 °C, 3rd step = 1 min to reach 210 °C, 4th step = 5 min at 210 °C, 5th step = 1 min to reach 220 °C, 6th step = 5 min at 220 °C and 7th step = 30 min to cool down to room temperature. The resultant solutions were transferred into 50 mL volumetric flasks using DI water. After the samples were filtered into plastic flasks the analysis of the mineral elements followed using ICP-OES (iCAP 6300 Duo MFC, Fa. Thermo; Waltham, USA). The operating power of the ICP-OES was 1150 W RF and the nebulizer gas flow 0.55 L min^{-1}. The carrier gas and plasmogen was argon and the cross-flow nebulizer (MIRA Mist; Thermo Scientific, Cambridge, UK) were used for performance for axial (Fe) and radial (P, K, Ca, Mg), view. To prepare a reference solutions for each element, a single-element solution were prepared in 1.4 mol L^{-1} HNO_3 and the calibration curves were created with following reference solutions: blank 1.4 mol L^{-1} HNO_3; 0–50 mg L^{-1} of P and Mg, and 0–10 mg L^{-1} of Na, 0–400 mg L^{-1} of K; 0–100 mg L^{-1} of Ca, 0–20 mg L^{-1} of S, 0–5 mg L^{-1} of Fe and Zn, and 0–1 mg L^{-1} of Mn, Mo, and B. The respective elements in the digestion solution were analysed at wavelengths describes in 8.3.2.1. The backup ensued by double determination and the results were expressed as mg 100 g^{-1} FW.

8.7.4 S1-4 Qualitative and quantitative analysis of phenolic compounds

The compounds were separated at 35 °C and a flow rate of 0.4 mL min.$^{-1}$ using a 150 x 2.1 mm, 3 µm, C16 column (AcclaimPA, Thermo Fisher Scientific; Waltham MA, USA). For analysis, eluent A (0.5% formic acid in water, Merck; Darmstadt, Germany) and eluent B (40% Acetonitril, J.T. Baker; Deventer, The Netherlands) were used. As such, following eluent gradient was used: 0–1 min 0.5% B, 1–10 min 0.5–40% B, 10–12 min 40% B, 12–18 min 40–80% B, 18–20 min 80% B, 20–24 min 80–99% B, 24–30 min 99–100% B, 30–34 min 100–0.5% B, and 34–39 min 0.5% B.

8.7.5 S1-5 Data used for the calculation of CO_2-equivalents

Fertilizer	Nutrient content of source (%)
CAN	15% N, 22% Ca
KNO_3	13% N, 38% K
$MgSO_4$	
KH_2PO_4	23% P, 28% K

Element	$CO_{2\text{-}e}$ per nutrient based fertilizer ($g\ g^{-1}$)
N	7.6
P	1.26
K	1.2
Ca	0.311
$MgSO_4$ (total fertilizer)	0.295

Example for calculation:

117 g CAN per L stock solution were used (control). 117 g CAN contains 17.6 g N and 25.7 g Ca. 17.6 g N result in 133.3 g $CO_{2\text{-}e}$ per (7.6 $g\ g^{-1}$), and 25.7 g Ca in 8 g $CO_{2\text{-}e}$ (0.311 $g\ g^{-1}$). In total the $CO_{2\text{-}e}$ for CAN in the control is 141.4 g.

9

General discussion

9 General discussion

Globally, an increase in food production is required; without simultaneously increasing the negative impact on the environment. This global challenge is called "sustainable intensification" (Godfray et al., 2010). Aquaponics is, in fact, a sustainable food production system, but the overall output of conventional low-tech aquaponics, SRAPS, is relatively low. As such, double recirculating aquaponic systems (DRAPS) are an approach to contribute to the challenges of feeding the increasing world population. The approach aims to intensify food production (fish and plants) while simultaneously keeping the idea of resource preservation and sustainability. For the implementation of DRAPS as a true alternative for conventional food production systems (aquaculture and horticulture), DRAPS has to compete with large, commercial and intensive fish and plant production systems. The basic concept of DRAPS can only be realised when optimised conditions are implemented for both productions units – fish and plants. Due to the independence of both production units (fish and plants), the critical point of DRAPS is obviously the connection point between the fish and plant unit, and the adjustment of the fish water for plant nutrition, respectively.

As mentioned in 2.3.2, the proof of concept regarding DRAPS is lacking until now. Besides the experiments of the present work, only four further empiric investigations were published dealing with DRAPS (Delaide et al., 2016; Goddek & Vermeulen, 2018; Kloas et al., 2015; Monsees et al., 2017b). Only two of them (Kloas et al., 2015; Monsees et al., 2017b) investigated DRAPS already as a whole. However, latter experiments were conducted in a prototype, where fish and plants were produced together in one spatial unit. As result, the hydroponic unit/number of plants was relatively small in comparison to the fish rearing unit and no comparison to conventional single hydroponic production was made. To make DRAPS competitive with conventional single hydroponic systems, a direct comparison between both systems is absolutely necessary. However, until now only the publications of Delaide et al. (2016) and Goddek and Vermeulen (2018) are known, which compared empirically the concept of DRAPS to conventional hydroponics. Goddek and Vermeulen (2018) however, used for their RAS based treatment only 30% fish water and 70% rainwater.

The general discussion of the present thesis will focus on four topics. Because the competitiveness to conventional hydroponic plant production is absolutely important, the 1st subsection will focus on plant production quality in DRAPS in comparison to conventional hydroponics. In the 2nd subsection, nutrient management will be discussed because the connection point between the fish rearing system is one important bottleneck in DRAPS. The 3rd and 4th subsections are used to discuss briefly the optimal fish to plant ratio and the contribution of DRAPS to environmental relief, respectively.

9.1 Plant production in double recirculating aquaponic systems – quantity and quality parameters

As discussed in chapter 8, Delaide et al. (2016) found that the FM of lettuce was enhanced by 39% in DRAPS in comparison to single hydroponic production or SRAPS. These results could not be confirmed by the investigations for the present thesis. In experiment 1 and 2, done in the INAPRO research DRAPS, the tomato fruit yields were equal in DRAPS and conventional single hydroponics (control, fresh water used to mix the nutrient solution) (chapter 3 and 4, Figure 4 and Figure 7). The yield results for tomatoes were independent whether the tomatoes were produced in combination with Nile tilapia (chapter 3 and 4) or African catfish (chapter 6). Additionally, in the separated experiment with lettuce (experiment 3, chapter 8), the nutrient solution prepared with fish water did not show an effect on plant growth compared to conventional single hydroponics. Interestingly, Goddek and Vermeulen (2018) recently published a similar study as Delaide et al. (2016) and confirmed the significantly increased yield in supplemented aquaponics (equates DRAPS). However, the measured fresh matter yield increments in supplemented aquaponics was only 8% as opposed to the 39% increase reported Delaide et al. (2016).

The general hypothesis of Delaide et al. (2016) was that the enhanced growth of lettuce in DRAPS is caused by humid acids or plant-promoting substances contained in the fish water. Goddek (2017) and Goddek and Vermeulen (2018) supposed that the increased plant growth in DRAPS may species-specific or is related to the bacterial communities in the related system. As already mentioned in chapter 8 many different microorganisms settle in RAS (mainly biofilter) and the microorganism community differs between individual RAS (Leonard et al., 2000; Schreier et al., 2010; Sugita et al., 2005). Recently, Eck et al. (2019) confirmed that the bacterial community differs in the different systems – especially at genius level. They stated that some of the identified taxa could have beneficial function for plant health and growth. Therefore, the different communities might be one explanation for better lettuce growth reported by Delaide et al. (2016) and the contradictory results in the present studies. However, in the experiments conducted for the present thesis different plant-fish combinations were used. In 2015 (experiment 1, chapter 3, 4) Nile tilapia and tomato were produced in the INAPRO research DRAPS. In the following year (experiment 2, chapter 6), African catfish and tomato were combined. However, for the separated lettuce experiment (experiment 3, chapter 8) fish waste water from Nile tilapia from an external RAS was used. The latter was obtained from Leibniz-Institute of Freshwater Ecology and Inland Fisheries and therefore different to that used in the INAPRO research DRAPS. In no case, independent of the respective fish-plant combination, was the plant growth increased compared to single hydroponics. In this context, Delaide (2017) noted in his doctoral thesis furthermore that not an increased nutrient concentration, but the adjustment of the nutrient ratios might play an important role to enhance plant growth. Unfortunately, a more detailed discussion is lacking. However, in the separated lettuce experiment conducted for the present thesis, the N:P and the N:S ratios differed clearly in the control (conventional hydroponic) and the aquaponic treatments (Table 24). In terms of the N:P ratio the aquaponic treatments were closer to the ratio in the original nutrient solution recipe, given by Hochmuth (2001). For the N:S ratio it was the other way round. The N ratio with other macronutrients had less clear differences. To evaluate the effect of the nutrient ratio, as suggested by Delaide (2017), further investigations are necessary. As such, an experiment with different nutrient ratios has to be conducted and would furthermore also apply for conventional hydroponics. The different research results underline that much more research on DRAPS must be conducted.

Table 24: Different macronutrients to nitrogen ratios (mean values ± standard deviation) in different nutrient solutions used in experiment 3 (chapter 8).

Element ratio	Control	AP*unt*	AP*dis*	Recommended by Hochmuth (2001)[b]
N_{min}[a]	205.4 ± 9.1	199.4 ± 10.5	193.5 ± 23.0	200
N:P	1.42 ± 0.06	2.72 ± 0.16	2.60 ± 0.33	3.23
N:K	1.39 ± 0.07	1.37 ± 0.07	1.40 ± 0.14	1.33
N:Ca	0.72 ± 0.04	0.63 ± 0.03	0.64 ± 0.07	0.95
N:Mg	3.95 ± 0.17	3.56 ± 0.20	3.58 ± 0.42	4.00
N:S	2.45 ± 0.15	1.45 ± 0.08	1.44 ± 0.17	2.86

Control = nutrient solution prepared with fresh water and mineral fertiliser.
AP*unt* = nutrient solution prepared with fish waste water.
AP*dis* = nutrient solution prepared with fish waste water; disinfected by heating (see chapter 8).
[a]Final mean N_{min} concentration in the used nutrient solution (see chapter 8, Table 20).
[b]Nutrient solution in experiment 3 was prepared according to Hochmuth (2001).

In chapter 5 it was demonstrated that the oxygen concentration in the cultivation trenches can drop temporary to critical levels when fish waste water was used. The measurements were done in the original DRAPS (incl. 3-chamber pit (3-cp)). However, the relatively low oxygen concentration in the nutrient solution did not affect the fruit quantity in the present study (chapter 5). But using the original DRAPS with a 3-cp requires a critical observation of the oxygen concentration in the nutrient solution prepared with fish water. Due to the comprehensive discussion in subsection 5.5, this wills not discussion further at this place.

The outcome of the different experiments is that it is possible to gain the same yields in DRAPS as in single hydroponics. This is an excellent result. The product profitability of greenhouse vegetable depends directly on the yield (Lattauschke, 2000). Therefore this parameter is absolutely important for producers and therefore also for the establishment of DRAPS as an alternative food production system.

Besides quantity aspects with economic importance for the producers, external and internal quality parameters of products play an important role for consumer decisions. The external quality (e.g. colour or firmness of tomato fruit) is, for instance, important when consumers buy foods (Causse et al., 2003), while the internal quality matters for health related aspects and taste (Wang et al., 2014). As such, to confirm that vegetable quality can be maintained using DRAPS, different quality parameters were examined for the present thesis.

It was demonstrated that the use of fish waste water did not influence the secondary metabolites lycopene and ß-carotene in tomato fruit when fish water from Nile tilapia (experiment 1) was used to mix the nutrient solution (subsection 3.4.2, Figure 5). Similar results were found in the following year (experiment 3), where fish waste water from rearing African catfish was used (Figure 23). Therefore, from the present point of view, it can be concluded that fruit quality in terms of the analysed secondary metabolites can be maintained when tomatoes were produced in DRAPS. This statement was further supported by the separate lettuce experiment results gained in experiment 3 (chapter 8). In terms of phenolic acids and flavonoids, no differences were found between the treatments (subsection 8.4.8, Table 23). However, when *Brassica* species are produced in DRAPS, attention should be drawn to glucosinolates. The accumulation of these metabolites can be affected by the sulphur concentration (Falk et al., 2007) in the nutrient solution. As shown in subsection 8.4.3 (Table 20), the sulphur concentration in the nutrient solution can differ (see also chapter 9.2) between conventional hydroponics and aquaponics. Until now only a few studies on aquaponics are known where the effects on secondary metabolites were evaluated. Besides this, the respective studies were only done in SRAPS. Schmautz et al. (2016) investigated the effect of different SRAPS on tomato fruit quality in comparison to conventional hydroponic systems. The SRAPS differed in terms of the hydroponic unit. They found a significant higher lycopene concentration in conventionally produced fruits. Roosta and Ghorbani (2011) investigated the essential oil content in *Mentha* species. While the essential oil content in *Mentha sativa* was increased by conventional hydroponics, it was increased by aquaponics in *M. piperita*.

In the 2015 study (Nile tilapia + tomato; experiment 1, chapter 3) the dry matter and soluble solid content (SSC), as well as the sugar-acid ratio (SAR) were significantly increased in hydroponics. However, these differences were not detected in the subsequent experiment performed one year after (African catfish + tomatoes; experiment 2, Figure 23). Aquaponic systems are highly complex. Therefore, the interpretation of the different results cannot be linked to one specific factor. Possible influencing factors are discussed in subchapter 3.5.2.

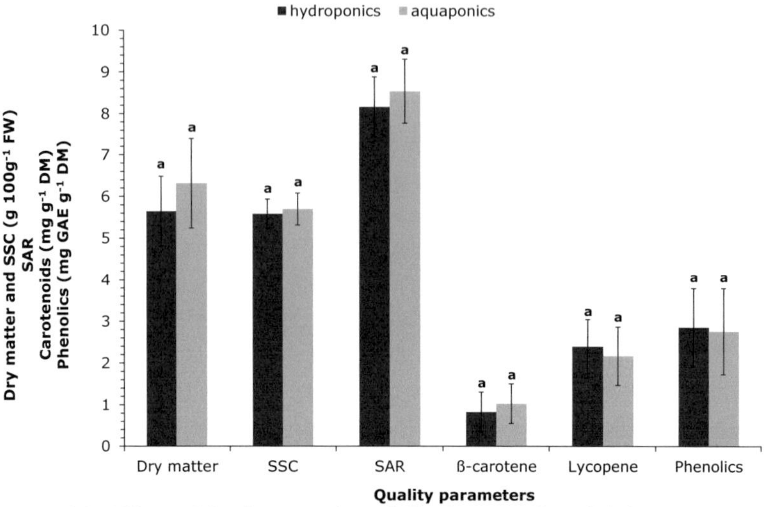

Figure 23: Effect of fresh water based (hydroponics) and fish waste water based (aquaponics) nutrient solutions on fruit dry matter, soluble solids content (SSC), and sugar-acid ratio (SAR), as well as ß-carotene, lycopene, and total phenol (phenolics) content. The fish waste water came from African catfish rearing (experiment 2). The data represent mean values and ± standard deviation (n = 9). The analysed contents were compared using t-test and represent the mean of three repetitions in three consecutive weeks. Different small letters indicate significant differences ($p < 0.05$).

The SSC and the SAR, are important factors for the overall flavour of tomato fruit (Baldwin et al., 1998; Stevens et al., 1979). In chapter 3 it was discussed that the fruit flavour might be affected negatively when produced in DRAPS. Indeed, additional data in Figure 24A presenting the results of a sensory test (hedonic test with untrained persons). It was found that fruit in aquaponics with a low EC (EC 1.8) were significantly less sweet and juicy, more bitter, had higher peel firmness and undesired aftertaste compared to conventional hydroponics (EC 1.8). Except for the undesired aftertaste, the differences did not apply to fruit produced in aquaponics with high EC (EC 3.0). However, the tomato-like taste did not differ between the treatments. The results gained in 2015 (experiment 1, chapter 3 and 4) were not confirmed by the experiment in the following year (experiment 2). Tomato fruit produced with fish waste water from African catfish were significant more tomato-like, less sour and bitter, juicier, as well as more fruity than conventionally produced tomato fruit (Figure 24B).

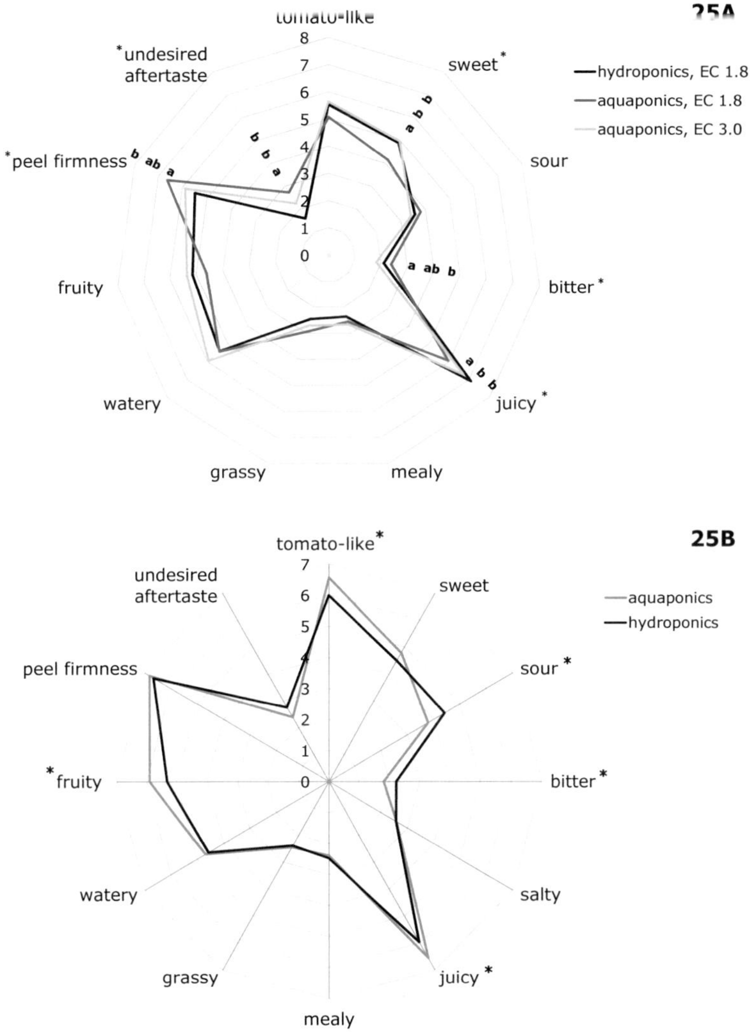

Figure 24: Sensory investigations regarding mouthfeel attributes and flavour. Tomatoes grown in fresh water (hydroponics) and two different fish waste water based nutrient solutions (aquaponics, 25A = EC 1.8 and EC 3.0). The data represent mean values from two different years and experiments (A = 2015, experiment 1; B = 2016, experiment 2). Values were compared using t-test (A: n = 55; B: n = 93) and different small letters and asterisks indicate significant differences ($p < 0.05$). The intensity of all sensory impressions is defined as 0 (low intensity) and 10 (high intensity).

Parallel to the second sensory test (Figure 24B) a laboratory test (Table 25) was conducted to analyse fruit SSC and SAR. In both cases, no significant differences were found. However, the fruit taste cannot only explained by the SAR and SSC. Flavouring substances also play an important role for tomato fruit taste (Herrmann, 1979). Additionally, it must be emphasised that the sensory tests were conducted as a hedonic test with untrained persons. While for the first test (experiment 1) 55 persons were surveyed, the second test (experiment 2) evaluated data from 93 persons. For hedonic tests usually a minimum of 100 test persons are necessary. Therefore the findings should be interpreted with reservation. Nevertheless, the results showed that fruits produced in DRAPS can differ in taste to those produced in conventional hydroponics. But the taste can be worse or better.

Table 25: Mean soluble solid content and sugar-acid ratio of tomato fruits grown in fresh water (hydroponics) and fish waste water based (aquaponics) nutrient solution.

Treatment	Soluble solid content ($g\ 100g^{-1}$ FW)	sugar-acid ratio
Hydroponics	5.26 ± 0.13^a	11.29 ± 0.58^a
Aquaponics	5.34 ± 0.08^a	11.31 ± 0.67^a

The mean data (± standard deviation) were compared using t-test and different superscript letters indicate significant differences ($p < 0.05$).
The data represent mean values ($n = 4$) of a laboratory analysis done parallel to the sensory test in 2016 (experiment 2, see Figure 24B).

The overall effect of DRAPS on fruit quality parameters cannot be clearly evaluated. While some parameters can be affected negatively (dry matter, SSC, SAR) in one year, the same attribute may be similar in the following year. Perhaps as suggested earlier, the fish species might play a role. But more studies are necessary for a conclusive evaluation. The quality of fish waste water is based on fish species, but also on fish rearing and operation management. All experiments of the present thesis were subject to different management conditions; therefore no general statement about fish management´s effects on plant quality can be made. In summary fruit quantity as well as the fruit quality can compete with conventional hydroponics. Furthermore, it can be stated that customers do not have to choose between sustainability and quality.

9.2 Nutrient management in double recirculating aquaponic systems

As demonstrated by the present thesis, in DRAPS the fish water can be optimised for plants, independent from its quality delivered by the RAS. This statement was confirmed by several authors (Goddek, 2017; Kloas et al., 2015; Monsees et al., 2017b). Therefore DRAPS can compete with conventional single hydroponic tomato production (see subsection 9.1). However, until now there is a lack of information regarding the connection point of DRAPS – between fish and plants. This has not been well investigated or discussed up to now, though some other authors have recently published their doctoral theses concerning aquaponics (Delaide, 2017; Goddek, 2017; Monsees, 2018). This chapter will provide information on this topic and shall help to give a general overview of the complexity of this topic.

Both experiments conducted in the INAPRO research DRAPS (experiment 1 in 2015 and experiment 2 in 2016) started from zero in terms of fish rearing. That means that fish stocking started in January in either case with low stocking densities in order to activate the biofilter. With increasing stocking densities a continuous adaption of the nutrient stock solutions[14] was necessary. Furthermore, in experiment 1, the Nile tilapia production was lower than expected. Consequently, the nutrient concentration in the fish waste water was relative low (especially N) in comparison to conventional Nile tilapia rearing RAS (Table 2) or to Nile tilapia reared in DRAPS by Kloas et al. (2015) and Monsees et al. (2017b). Even with the suboptimal fish production, the plant production could be kept at high performance. This result emphasises the clear advantage of the DRAPS where both units can be optimised individually. Furthermore, a complete disconnect of both systems would be possible when a problem occurs in one of the systems. This would prevent a total collapse of the system and the unconcerned production unit can be saved. The possibility of adjustment was further demonstrated in the second experiment, conducted in 2016 (chapter 6). Here, a modification of the operating DRAPS system occurred and required a comprehensive adaption of the nutrient stock solutions. Additionally, it was found that the nutrient concentration in the fish water (sampled in the RAS), can even fluctuate in fully stocked RAS as demonstrated in Figure 25. The fluctuations in fish water from Nile tilapia (Figure 25A) was stronger

[14]Two concentrated stock solutions were mixed to prepare the nutrient solution for hydroponics. See subsection 6.3.2.2, Figure 14.

than in fish water from African catfish (Figure 25B). The stronger oscillations in fish water from Nile tilapia was presumably primarily a result of the relatively low performance and growth of the fish. However, the graphs in Figure 25 represent results of one RAS, where two samples were taken at each measuring day. Therefore, these results should be interpreted with care.

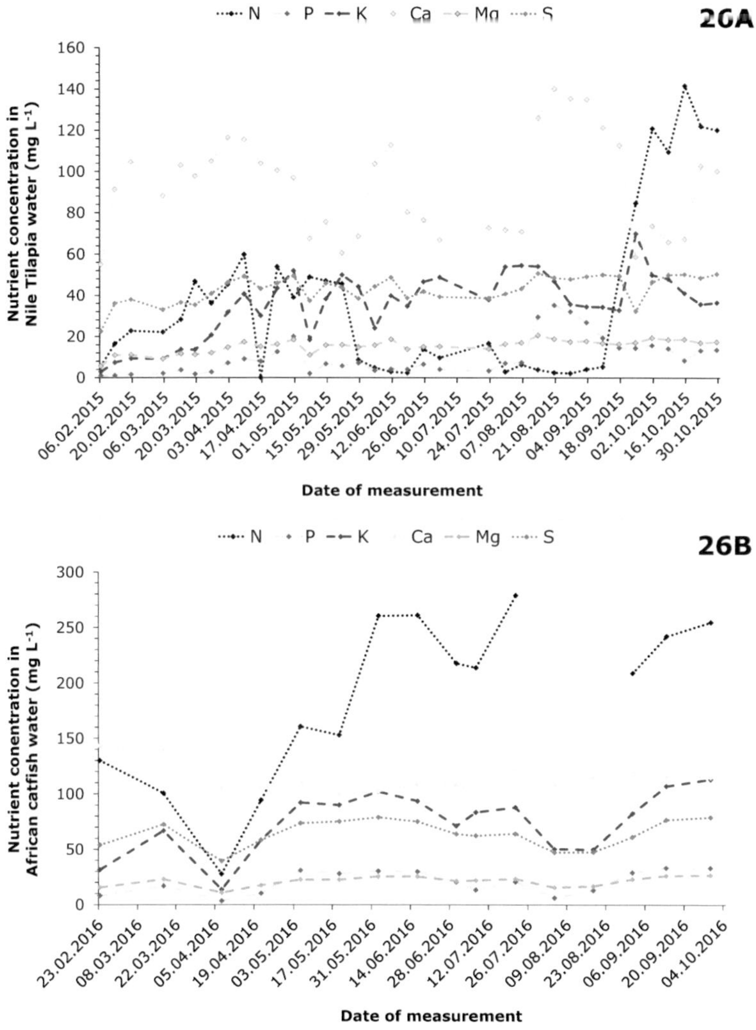

Figure 25: Macronutrient concentrations in fish water (sampled in one RAS) measured in Nile tilapia water (2015, 26A) and in African catfish water (2016, 26B) on a weekly (26A) and two-weekly basis (26B), respectively (n = 2). RAS fully stocked at 31.05.2016.

Due to fluctuating nutrient concentration in the fish water a continuous measurement of the fish water and adjustment of the stock solutions would be necessary. A continuous measurement and adjustment of the stock solutions is possible for experimental evaluations (see below); currently this is not an option for a commercial operation of DRAPS. Analysis are usually very expensive and a regular adjustment is additional highly time-consuming[15]. Therefore, an online real-time measurement system with direct feedback control and single nutrient tanks for each mineral salt would be a meaningful contribution to DRAPS (see chapter 3). This would not only be helpful to respond to changing nutrient concentrations in fish waste water, but also as an enhanced nutrient management system in the recirculating (closed) hydroponic unit. The latter becomes also a more and more important issue for the conventional recirculating hydroponic systems (Bamsey et al., 2012; Jung et al., 2015). However, up until now no system has been is established in the commercial market.

An alternative approach to enhance the adjustment of the fish waste water could be the stabilisation of the nutrient concentration in the fish water itself. With experiment 2 (measured in 2017, chapter 7) it was shown that the mechanical filter system might have a significant impact on nitrogen stability. In 2016 the sedimentation unit (see subsection 6.3.1, Figure 13) was replaced by a new suction filter device (described in chapter 6 and 7). Over a period of three consecutive days in September 2016 the nitrogen concentration remained relatively stable between 253 mg L^{-1} and 283 mg L^{-1} (263.4 mg L^{-1} ± 6.7; chapter 7, Figure 18). In 2017 a repetition of the measurement was performed after the first prototype I was replaced by the enhanced prototype II (chapter 7, Table 16). But in 2017 the measurements were done over a period of 15 consecutive days. An initial evaluation was unable to convincingly confirm the results gained in 2016. Over the measuring period of 15 days a slight, but consistent increase in the nitrogen concentration was detected (from 163 to 263 mg L^{-1}). Beside the effect on nitrogen, the macro- and micronutrient concentrations could be kept also relatively stable (Figure 26 and Figure 27, respectively). The higher K concentration measured in the storage tank was due to the supplementation of potassium carbonate ($KHCO_3$) and potassium hydroxide (KOH). This was supplemented to storage tank because in the fish water, delivered from African catfish, high amounts of ammonium nitrogen were detected. The addition of carbonate and lye increased the pH value of the fish water which in turn supports the nitrification processes, converting ammonium to nitrate. The ammonium issue was already discussed in subsection 6.4.2. The decreased nutrient concentrations in the middle of the measuring period could

[15]Experience of practise partners in the INAPRO project

not be explained. Maybe something happened in the fish rearing management (water loss, lower feed) which was not recorded. The fluctuation of the micronutrients B and Mn can also not be explained. Interestingly, the other micronutrients did not follow the course.

Using the new filter device resulted in a continuous removal of sludge from the RAS and a regular water transfer from the RAS to the storage tank (Figure 13b). The water was transferred several times a day in small amounts (e.g. 4 x 50 L) from the RAS into the storage tank. As already mentioned in subsection 7.4.2, this could be responsible for the relatively stable nutrient concentrations. In comparison to the use of the sedimentation unit (original DRAPS, Figure 13a), only a very small amount of water (50 L ≙ 0.4% of the total RAS water volume) was replaced at one moment. When the sedimentation filter, with a water volume of 1.3 m^3 is cleaned, around 10% of the whole RAS water volume is replaced at one moment.

Besides the effect on nutrient stability, the direct transfer of the fish water to the storage tank, without passing the 3-cp, resulted in immense reduced nitrogen loss. The nitrogen loss occurred in the 3-cp, probably due to denitrification processes within the first chamber. This was discussed in detail in subsection 6.4.3.

Unfortunately, the influence of the new filter device on the nutrient concentrations was just based on first investigations without repetitions. Therefore the discussions made in this context were just an assumption. However, the results gained by the new suction device in regards to nutrient stability were very promising. Further investigations are highly recommended. To summarise, it could be demonstrated that the adjustment of the nutrient solution is possible, continuous small volume water exchange probably stabilises the nutrient concentrations in fish (waste) water and nutrient loss can be minimised. However, the continuous and automatic control of the nutrient concentration will be an absolute necessary tool for the practical use of DRAPS.

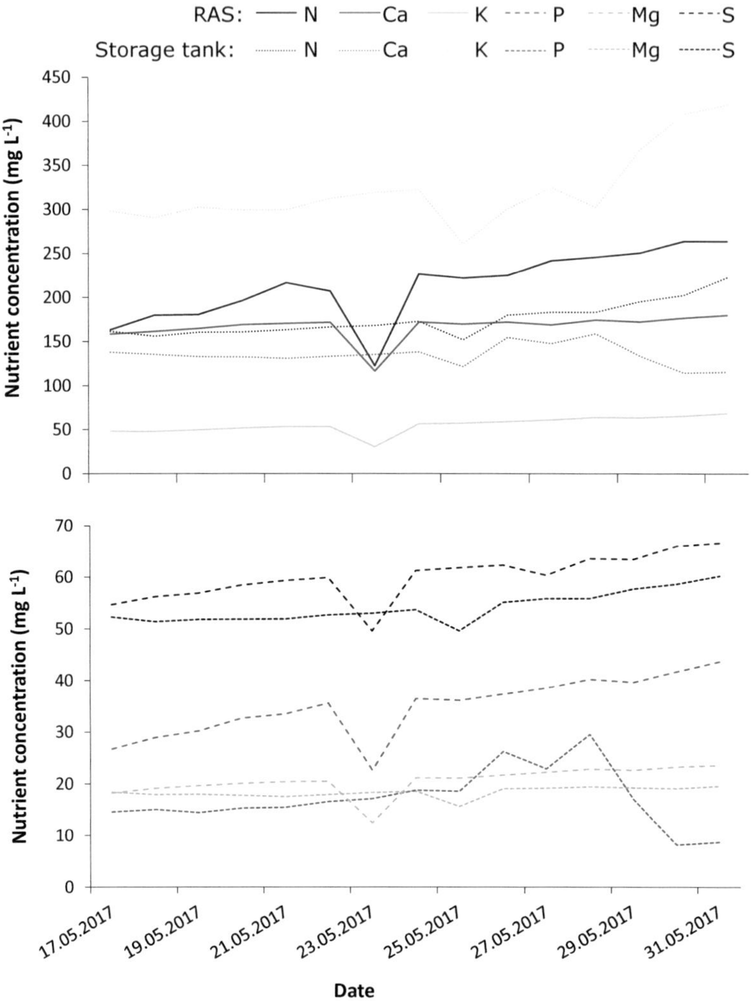

Figure 26: Macronutrient concentration measured on 15 consecutive days in May 2017 within the recirculating aquaculture system rearing African catfish (solid line) and storage tank (dashed line) in the modified double recirculating aquaponic system (see chapter 6).

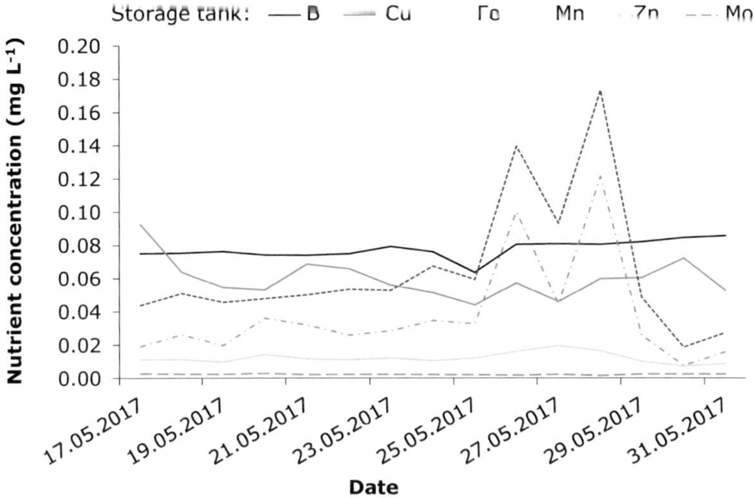

Figure 27: Micronutrient concentration measured on 15 consecutive days in May 2017 within the storage tank in the modified DRAPS (see chapter 6).

In experiment 3 the nutrient solutions were tested directly after preparing. Therewith it was possible to evaluate the effective nutrient concentration and to visualise differences to conventional hydroponics. As demonstrated in chapter 8 (experiment 3), the exact adaption of the fish waste water to the recommended nutrient concentrations was not completely possible because multi-component fertiliser was used. Delaide et al. (2016) came to equal finding and could not realise completely identical nutrient concentration in DRAPS and the control. This is also an issue in conventional hydroponics because the nutrient concentration in the used fresh water (well water/tap water) must be considered. Table 26 presents the final calculated macronutrient concentration of different nutrient solutions. They were calculated for the different fish waste waters presented in Table 2. For comparison, the recommended nutrient concentrations for tomatoes (Lattauschke, 2004a) and the calculated concentration for conventional hydroponics are given. For calculation, conventional mineral fertiliser was used[16] and for hydroponics, mixed process water with tap water:rainwater

[16] $Ca(NO_3)_2$, KNO_3, NH_4NO_3 (ensured that in the final solution NH_4-N did not exceed 30%), K_2SO_4, K_2SO_4+Mg, $MgSO_4$, KH_2PO_4, $KHCO_3$

(50/50; v/v)[17] was utilised (also used in experiment 3, described in chapter 8). For calculation, a K:Ca ratio of 1.8 was considered (like in the recommended nutrient concentration).

Table 26: Calculated, theoretical final macronutrient concentration in nutrient solution based on different fish waste waters, delivered by three different fish species (Nile tilapia, perch, and African catfish) in comparison to fresh water based nutrient solution as in conventional hydroponics (tap water:rainwater; v/v = 50/50).

	Final macronutrient concentration							
	N	**P**	**K**	**Ca**	**Mg**	**S**	**Total**[a]	**Reduction**[b]
			mg L^{-1}				**g L^{-1}**	**%**
Recommended for tomatoes[c]	151.2	37.2	234.6	128.3	24.3	65.0		
Tap water:rain water (50/50; v/v)	151.3	37.3	314.7	176.8	25.0	69.2	147.2	
Nile Tilapia	151.8	37.2	170.6	93.2	24.9	63.6	52.0	64.7
Perch	151.3	37.2	229.5	126.1	28.2	60.0	98.4	33.2
African catfish[d]	225.6	37.4	204.7	112.1	28.8	62.3	20.2	86.3

For calculation, the mean nutrient concentrations (arithmetic average) in fish water displayed in Table 2 were used.
[a]Total amount of added fertiliser (excl. micronutrients) in g per 1 L stock solution.
[b]Reduction of total mineral fertiliser supply in comparison to tap water:rainwater (= 100%).
[c]By Lattauschke (2004a).

Due to the high calcium content in the tap water:rain water (50/50; v/v), the K and Ca concentration in conventional hydroponics was 1.3-fold and 1.4-fold higher, respectively, than suggested. Considering the final nutrient solution the best results were obtained using water from rearing perch. This was due to the lowest nutrient concentrations what provided more scope for preparing. But, simultaneously this also means that perch water results in the lowest mineral fertiliser use reduction. However, it is clearly visible that especially the K and Ca concentrations differ between the calculated nutrient solutions. Both nutrients are primarily bonded on nitrate as calcium nitrate ($Ca(NO_3)_2$) and potassium nitrate (KNO_3). Though potassium salt is also available in further combinations – as potassium (magnesium) sulphate ($K_2SO_4(+Mg)$), monopotassium phosphate (KH_2PO_4) and potassium chloride

[17]Data from Humboldt-Universität zu Berlin, Campus Dahlem were used for calculation

(KCl) (Hochmuth, 2001; Snyder, 2016). For fertiliser used in hydroponics, solubility is highly important and fertiliser must be dissolved completely (Snyder, 2016). In general, most calcium salts are scarcely or almost not soluble in water. Therefore, only $Ca(NO_3)_2$ and calcium chloride ($CaCl_2$) are suggested as calcium fertiliser (Hochmuth, 2001; Snyder, 2016). In recirculating hydroponic systems, sodium and chloride tend to accumulate and can cause salt stress (Savvas et al., 2007). To prevent salt accumulation in form of chloride, CaCl and KCl were not considered for preparation and therefore also here not for calculation. Especially with high N and low Ca concentrations in the process water (fish waste water) the addition of Ca is difficult (Nile tilapia, Table 2 & Table 26) and results in a Ca concentration below the recommendations. This limitation applied for both Nile tilapia and African catfish. Due to the higher Ca concentration in the water of African catfish, the Ca and K concentration were closer to the recommendations. Additionally, to raise the concentration of K, KOH was used in the calculation. This is not a generally used fertiliser. When used as fertiliser it must be considered that KOH has an alkaline effect. This might operate against the lowering of the pH value of the nutrient solution for hydroponics using acid. Depending on the fresh water quality it is common practise to lime RAS water to stabilise alkalinity and hardness. Well established materials used in RAS are hydrogen carbonates (HCO_3^-; Ca~, Na~, Mg~), hydroxide (OH^-; Na~, Ca~, Mg~) or calcium oxide (CaO) (Timmons et al., 2010). Therefore it might be possible to add the KOH not only at the mixing station of the DRAPS for preparing the nutrient solution but already in the RAS (partly). If necessary, also in combination with Ca supplements. This would reduce the difficulties of adjusting the fish waste water in terms of Ca and K. The addition of KOH or K_2CO_3 is, for instance, is often used in SRAPS and seems to be appropriate for supplementation of K (Rakocy et al., 2006; Somerville et al., 2014). However, an explicit experiment studying the effect of different alkalinity materials on fish welfare is missing and should be investigated.

In addition, the nitrogen concentration is an issue in itself. As shown in Table 2 and Table 26, the nitrogen concentration in fish water, e.g. African catfish, can reach levels above the recommended nutrient concentrations for plants. However, it must be considered that the recommend nitrogen concentrations for tomatoes differs and can be higher than those presented in Table 26 (see below and Table 27). On the one hand, high nitrogen concentrations can be problematic for the adjustment of nutrients as discussed above. On the other hand, nitrogen concentration above threshold concentration can cause phytosanitary problems during the cultivation and should also be considered. For instance, high nitrogen concentrations (308 mg L^{-1}) can result in thinner leaves and enhances the development of the greenhouse whitefly population (Jauset et al., 2000). Furthermore, it can make tomatoes more attractive for two-spotted spider mites (Hoffland et al., 2000a). To adjust the right nitrogen concentration two different approaches are conceivable. On the one hand, the adjustment can start in the RAS with its management (water exchange rate, stocking density). Alternately (fresh) water can be used to dilute the fish water before it is used for plant irrigation. The latter does not follow the idea of sustainability and water saving when tap water is used. But using rainwater, collected from the roof area, or the regained evapotranspiration water by an active cooling system in the greenhouse (see chapter 6.2; DRAPS full version - see Nomenclature and descriptions) could be used to meet the sustainable approach of DRAPS.

As discussed above, nutrient adaption is not as easy as implied by the system-level description of DRAPS. Less is reported comprehensively about real nutrient concentrations and their fluctuations in RAS. Certainly, the fluctuation might depend on various things like fresh water source, fish feed, fish growth, and management practises. However, in the literature, a wide range of recommended nutrient concentrations can be found for the same plant species. Table 27 gives an overview, how the recommended nutrient concentrations differ for tomatoes. For example, the recommended nitrogen concentration varies between 133 and 241 mg L^{-1}. Also the other nutrients vary partly clearly between the different recipes. This implies that tomatoes can in general handle a wide range of nutrient concentrations. Furthermore, the nitrogen concentration in fish water from African catfish did not reach harmful concentrations considering the recommendations of De Kreij et al. (1999) (Table 27). However, usually, the nutrient solution for tomatoes is adjusted during the cultivations period according to the plants´ developmental state (De Kreij et al., 1999; Hochmuth, 2001; Lattauschke, 2004a; Peet & Welles, 2005; Snyder, 2016). In the present experiments, this ongoing adjustment of the nutrient solution was abandoned to simplify the complex experiments. Certainly, for conventional tomato production, the adjustment would be highly recommended.

Table 27: Recommended nutrient concentration for greenhouse tomato production in mg L^{-1}

According to[a]	--	**Jones (1983)**	--	**Modified Steiner solution by Larsen**	**Hochmuth & Hochmuth (1995)**	**OFRAM (2001)**
Nitrogen				171	150	210
NO_3-N	144.2	133	224.2			
NH_4-N	7		16.8			22
P	37.2	62	46.5	48	50	50
K	234.6	199	371.5	304	200	420
Ca	128.3	130	216.5	180	150	190
Mg	24.3	50	58.3	48	50	75
SO_4	144.2		446.9			
S		70			60	120
Fe	0.8	2.5	0.8	3	2.8	0.8
Mn	0.3	0.6	0.5	1–2	0.8	0.55
Mo	0.05	0.03	0.05	0.1	0.05	0.05
Zn	0.3	0.09	0.3	0.4	0.3	0.33
B	0.3	0.44	0.3	1	0.7	0.5
Cu	0.05	0.05	0.05	0.2	0.2	0.05
References	Lattauschke (2004a)[b]	Hochmuth (2001)[c]	De Kreij et al. (1999)[d]	Snyder (2016)[e]	Peet and Welles (2005)[c]	Peet and Welles (2005)[c]

[a]Given by the references, no access to primary literature.
[b]12th cluster to termination.
[c]Fifth fruit/cluster to termination.
[d]Tenth fruit to termination.
[e]100%; spring crop, above the fifth cluster until May.

9.3 Optimal fish to plant ratio

In SRAPS the fish to plant ratio is calculated as daily feed input to plant growth area. If the daily feed input is too high, nutrients accumulate. Alternately, when the daily feed input is too low, the plants suffer from a nutrient deficit (Rakocy et al., 2006). In the present studies, the calculation was different because the objective was the maximisation of water savings. Water saving was one main goal of the INAPRO project. Therefore the fish to plant ratio was based on the water amount delivered by the fish rearing system. However, model-based calculations also considered nutrient, especially nitrogen, balance (Reyes Lastiri et al., 2016).

The greenhouse area of the research INAPRO DRAPS (used for experiment 1 and 2) was too small in relation to the RAS. Only a small amount of the total fish waste water was used for plant irrigation (8.5% in 2015, chapter 3 and 15.3% in 2016, chapter 6). Based on maximal water reuse, it was calculated that 1 m^3 net fish production volume would deliver enough water for 28.2 m^2 to 34.3 m^2 net-acreage tomato production in greenhouses. Certainly, these data were only roughly calculated ratios. In this context, the calculation was based on fresh water use by the RAS and did not consider the evaporation in the RAS nor the regular nutrient solution replacement in the hydroponic unit.

The optimal fish to plant ratio is complex and depends on various factors. For instance, a higher or lower daily water exchange rate in the RAS would increase or decrease the waste water amount delivered by the RAS, respectively. Additionally, the water used by the plants and perhaps required dilution of the fish waste water, due to high nitrogen concentrations (see 9.2) would also affect the optimal ratio. The water use by tomatoes varies depending on the development stage and transpiration throughout the production period (from 0.2 to 3.2 L $plant^{-1}$ day^{-1}). The latter has a peak during summer time (Lattauschke, 2004a). Due to its complexity, the ratio is commonly calculated by empirical data (Goddek et al., 2016a) but optimal ratios for DRAPS have not been reported. For better understanding of the influencing factors, more specific studies on this topic must be carried out.

In the present studies, only the period of tomato production (January to October) was considered to calculate the ratio between fish and plant production unit. Fish rearing in RAS usually runs as continuous production strategy; tanks are stocked and harvested periodically the whole year round. In contrast, tomato production usually is interrupted by breaks in the summer or winter for pest control, avert unfavourable price seasons or to save cost for cooling, heating or lighting (Peet & Welles, 2005). Growers in southern latitudes like Spain, usually make a short break in the summertime. Alternatly, in northern latitudes like in the Netherlands and Germany, a break during winter time is common. For half year production systems short breaks during midwinter and midsummer are usual (Lattauschke, 2004a; Peet & Welles, 2005). As a consequence, the RAS produces waste water continuously year round, but the demand by plants varies strongly due to the season.

In comparison to intensive fish production in RAS, the avoidance of waste water disposal is an important advantage of aquaponics. The cost for nutrient-rich waste water (produced by RAS) can be very expensive (depending on location). For example, in a community (Waren) in Mecklenburg-Western Pomerania, Germany, the cost of fresh water is 2.75 € m^{-3}. The fee for "common" waste water is 2.69 € m^{-3} but for high amounts and highly polluted waters, a supplement is charged (Müritz-Wasser-/Abwasserzweckverband, 2011). Finally, the fee can rise to 4.50 € m^{-3} easily[18].

To minimise supply interruptions in summer time (Peet & Welles, 2005) and to increase the yield (Ávila-Juárez et al., 2012), some tomato growers work with intercropping/interplanting systems where young tomato plants are planted in the still existing "old" crop. Less is known about intercropping of different species in intensive, soilless greenhouse production. Mackowiak et al. (1999), for instance, examined the successful polyculture of soybean and tomato in closed hydroponic for NASA´s advanced life support. For intercropping tomato and lettuce, increased land use efficiency (Cecílio Filho et al., 2011; Jett et al., 2005) and improved economic indices (Cecílio Filho et al., 2010) were reported. In chapter 8 it was demonstrated that lettuce can be grown successfully in DRAPS. Due to its relatively low-temperature demand and fast growth (ready after 8 to 11 weeks) (Lattauschke, 2004b), lettuce would be well suited as an intercropping culture to bridge the winter break in greenhouses. Intercropping would help to make a part of the fish waste water delivered in winter time profitable.

[18]Verbal information by Ulf Rehberg, Fischerei Müritz-Plau GmbH

9.4 Environmental relief by double recirculating aquaponic systems

In general, aquaponics relieves the environment by using the nutrients in fish feed twice. Firstly as feed for fish and again as fertiliser in hydroponics.

In comparison to conventional hydroponics, the demand for mineral fertiliser is drastically reduced in DRAPS as demonstrated in chapter 6 and 8. For the present studies, the calculation was mainly based on nitrogen fertiliser (chapter 6 and 8). The focusing on nitrogen was mainly due to the relatively good data available for this type of fertiliser. However, in chapter 8 the other fertilisers used were also considered using ProBas database (UBA, 2015b). Certainly, the values for the present studies were calculated only on a rough basis. Nevertheless, one of the main environmental burdens of tomato production (Almeria, multi-tunnel greenhouse) is mineral fertilisers. It contributes mainly to eutrophication and global warming (Torrellas et al., 2012). In experiment 3 (chapter 8), it was found that 100% of the nitrogen demand of tomatoes can be delivered by African catfish reared in RAS. In the present study (experiment 1) this was not the case for Nile tilapia (chapter 3). However, this was due to the relatively low performance of the fish. As shown in Table 2 it is absolutely feasible that Nile tilapia can meet the N requirements of tomatoes when they are produced in RAS under professional conditions.

It was calculated that around 13 t $CO_{2\text{-}e}$ per ha tomato production can be saved when 100% N is delivered by fish (subsection 6.4.5). Extrapolated to the protected tomato production in Germany (374 ha; (AMI, 2017)) that means that 4 862 t $CO_{2\text{-}e}$ could be saved per year. For lettuce (chapter 8) equal amounts can be saved when lettuce is cultivated hydroponically on a yearly basis. However, the greenhouse cultivation area for tomatoes in Germany is relatively low. In The Netherlands, the 4th biggest tomato producer in the European Union (AMI, 2017), tomatoes are commonly produced in glass greenhouses in hydroponic systems. There, the protected total area amounts more than 9 850 ha (Peet & Welles, 2005). This, in turn, would result in saving 128 050 t $CO_{2\text{-}e}$ per year, 26-fold the amount of Germany. The reduction in $CO_{2\text{-}e}$ is a result primarily of the reduced manufacturing of N fertiliser. The production of this fertiliser is based on the Haber-Bosch process which is a very energy demanding process using natural gas (Wood & Cowie, 2004). This, in turn, means that reducing the demand of N fertiliser preserves fossil fuels, which is a further benefit. High amounts of nitrogen in the fish water are generally good for contributing to environmental relief. But, as already discussed in subsection 9.2; high nitrogen concentration in the fish water can prohibit the addition of Ca or K fertiliser. The calculated savings refer only to the hydroponic unit and only to the non-use of nitrogen fertilizer in aquaponics. This considers just a small part of DRAPS. For a better understanding of DRAPS, a complete life cycle assessment for the whole DRAPS should be made.

An important improvement of the research DRAPS was the reduction of nitrogen loss by the implementation of the new filter device (chapter 6 and 7). Due to leaving out the 3-cp, significant amounts of nitrogen and nutrients could be saved (chapter 6). This was a highly important development because the sustainability of the original DRAPS was drastically reduced by the loss of nitrogen. Nitrogen loss does not correspond with the idea of aquaponics; whether constructed as one or two loop system. The conservation of nutrients was also a particular issue to think double recirculating aquaponic one step further. As such Goddek et al. (2016a) presented a multi-loop aquaponic system. In addition to the fish and plant unit they integrated a sludge mineralisation component as a third loop. They regained minerals from the sludge which were subsequently delivered to the hydroponic unit. As already mentioned in subsection 2.3.2, Goddek (2017) promoted an anaerobic sludge mineralisation. In contrast, Monsees et al. (2017a) recommended an aerobic sludge mineralisation in order to prevent nitrogen loss by denitrification. In his doctoral thesis, Monsees (2018) remarked further that aerobic systems are easy to handle. It would be much easier to implement aerobic systems in DRAPS than highly complex anaerobic systems. Under practical conditions, increasing complexity probably raises in turn the risk of mismanagement which can waste resources, increase the workload or cause a drop out in one unit. In the worst case, it can finally result in a breach of trust in the whole double recirculating system. Therefore, it seems appropriate that the implementation of DRAPS as commercial working large-scale system should be performed step by step.

10

General conclusion and recommendations

10 General conclusion and recommendations

To come back to the formulated objectives in subsection 1.2.2 it can be concluded that the combined production of tomatoes and lettuce in DRAPS is as successful as in conventional hydroponics. Not only the quantity but also the quality of plants (tomato fruits and lettuce) can compete to that produced under conventional single hydroponic conditions. These results were confirmed in all conducted experiments, whether Nile tilapia or African catfish were reared. However, in contrast to other publications no increases in plant growth were observed when plants were irrigated with fish waste water as compared to conventional hydroponics.

The oxygen concentration in the cultivation trenches can drop to critical levels when the nutrient solution is prepared with fish waste water. However, no negative effects on plant performance were detected in the present study but oxygen levels should be generally controlled. Probably, the strong oxygen depletion/consumption was also related to the 3-cp, used in the original INAPRO research DRAPS. The oxygen consumption of the fish waste water which passed the 3-cp was noticeably high. The oxygen consumption was much higher than in the fish water sampled directly in the RAS. The modification of the original INAPRO research DRAPS might have improved the oxygen situation in the cultivation trenches, but was not studied further.

It was demonstrated that using DRAPS reduced the environmental impact due to the reduced demand for mineral fertiliser. Mineral fertiliser manufacturing, especially the production of nitrogen fertiliser is a high energy demanding process. Consequently, just due to the reduction in mineral fertiliser that must be manufactured, the environment can be preserved. The modification of the INAPRO research DRAPS by implementing a new suction filter device especially increased the potential for fertiliser savings and environmental relief. The new suction filter device which replaced the sedimentation unit reduced the nitrogen loss, occurred in the 3-cp, immensely. Unfortunately, a comprehensive study of the new suction filter device and its effects was not possible. Only first approaches could be done. But these results were very promising and can be an important component to improve the connection point between fish and plants. The disinfection of the nutrient solution prepared with fish waste water, and therefore the microorganism contained in the fish waste water, did not affect lettuce growth.

10.1 Summary of advantages and disadvantages of double recirculating aquaponic systems

To summarise briefly, the advantages and disadvantages of DRAPS the instrument of a simple SWOT (strengths, weaknesses, opportunities, and threats) analysis were used. The SWOT analysis is an analytic method of strategic corporate planning but is here used more generally to give a condensed overview of the advantages and disadvantages of DRAPS (Table 28). The strengths and weaknesses are internally-related. The opportunities and threats focus on the external environment of the system. With the present thesis primarily the strengths, but also some weaknesses of DRAPS could be identified and were discussed above in the different chapters. But to give a comprehensive overview not only the results of the present experiments, but also the experience gained during the four project years of INAPRO will be considered in the next part. Table 28 gives a brief summary of the strengths, weaknesses, opportunities, and threats described in following subsections.

Table 28: Overview of strengths, weaknesses, opportunities, and threats of double recirculating aquaponic systems.

Strengths	Weaknesses
• Independence of fish and plant units • Optimal conditions for each cultivated species • Able to compete with conventional hydroponics • Decoupling possible • Possibility to combine already existing systems • Sustainable food production system • High potential for water saving[1] • High potential for fertiliser saving • Reduced environmental impact • Space efficient • Meet increasing demand on fish	• Combination of two high tech systems • High investment costs • No economic evaluation • Minimal experience in (large-scale) production units • Know-how is lacking • Insufficient education • Know how/qualified staff in both parts is needed • Automated adjustment of the fish waste water currently not feasible • High control expenditure • Fish feed currently not sustainable • Personal responsibility not clear (independent or consolidated)
Opportunities	**Threats**
• Increasing demand for sustainable/local food production • Sustainable fish feed production from insect meal • Appreciation by consumers • Higher sales prices – by valuing consumers or retail/when recognised by consumers or retail • Higher sales prices by direct marketing • Answer for "sustainable intensification"[2] in food production systems	• Aquaponics is unknown • Unacknowledged/rejected by consumers • Environmental benefit not recognised/appreciated by consumers • Not recognised need • Hydroponic world not yet ready – old manners are the best proven

[1]Especially when the full version of INAPRO DRAPS is used.
[2]"Producing more food from the same area of land while reducing the environmental impacts" by Godfray et al. (2010).

10.1.1 Strengths

i. Production issues

As demonstrated in the present thesis (Chapter 3, 4, 6, and 8) and reported by other authors (Delaide et al., 2016; Monsees et al., 2017b) one primarily strength of DRAPS is the competitiveness to single fish or plant production in terms of both, quantity and quality aspects. This benefit is based on the independent production of fish and plants under respective optimal conditions. This independence also covers the system from a total blackout because decoupling is possible e.g. in times of damage of one system or other problems (pests, pest control, reduced growth) in one of the systems (chapter 3).

Another aspect is the modularity of the system which allows already existing systems to can be connected. To construct a DRAPS it is not necessary to build a new and complete unit. It is also necessary to combine already existing systems, which has been demonstrated by the INAPRO partner inagro (www.inagro.be). In that case, high investment cost can be reduced (see 10.1.2, weaknesses). However, the connection of two systems can also be problematic, when two different operators have to cooperate (see 10.1.4, threats).

ii. Sustainable issues

DRAPS were developed to find an answer for sustainable but intensive food production. Its sustainability is distinguished by saving scarce resources like water and nutrients. The double use of water and nutrients reduces the demand for fresh water and mineral fertiliser in comparison to single production systems. Beside the double use of water the full version of DRAPS (see Nomenclature and descriptions) can especially minimise the fresh water demand of the whole system. Using the fish feed as the nutrient source for fish and plants, minimises the demand of mineral fertiliser for plant fertilization. This, in turn, reduces drastically the emissions result from manufacturing of mineral fertiliser (chapter 6, 8, and subsection 9.4). Therewith DRAPS can contribute to environmental protection. Additionally, intensive indoor-farming is space-efficient and enables high productivity per area. That means on the one hand that DRAPS can meet the challenge for increasing demand of food and on the other hand that valuable space is optimally used.

Another issue is that any increasing fish demand will be met mainly by aquaculture (subsection 2.2). Therewith, DRAPS can contribute to health food supply for rising populations.

10.1.2 Weaknesses

i. Investment issues

DRAPS were in general developed as an intensive food production system. Therefore one identified weakness is that DRAPS combine two high tech systems – the RAS and the hydroponic system. The latter becomes high tech especially when the full version of DRAPS (subsection 6.2) is used. This, in turn, results in high investment costs, whereby an economic evaluation of DRAPS is currently lacking. However, the economic efficiency is highly important to make DRAPS interesting for producers and for the implementation as a commercial food production system.

ii. Know-how

Due to its relatively recent development and missing experience in large-scale production units, the know-how for successful operation of DRAPS is lacking. Due to the production of two totally different species, expertise in both subjects is necessary. Neither fish nor vegetable is only produced incidentally; therefor qualified staff is indispensable. However, until now, no specific education is offered to teach personnel comprehensively. But highly committed labour is needed to operate DRAPS approximately successfully. However, first experiences in operating commercial DRAPS were gained in the INAPRO demonstration facilities (Spain and Germany). With nearly 550 m^2 (15 t fish production in RAS + 288 m^2 net acreage hydroponic production) these demonstration facilities were larger than the already known DRAPS. But they are not so big as to reflect commercial large-scale greenhouses or fish production systems. For comparison, one of the biggest greenhouses in Germany has a size of around 15 ha and will grow to around 40 ha in 2020 (Lähne, 2016). A known indoor RAS for African catfish production produces around 120 t fish per year (bbv, 2014).

iii. Connection point between fish and plant

One main bottleneck is the connection point between the fish and plant unit. At the connection point, the fish waste water is optimised by adding mineral fertiliser and e.g. acid for optimal plant growth. An automatic adjustment would be necessary to make the system usable for commercial production. The regular control and adjustment of the nutrient stock solution is time, labour and cost consuming. But currently, no ready to use online real-time measurement system is available at the market (subsection 9.2).

iv. Fish feed

DRAPS were developed as a sustainable food production system. However, most of the conventional fish feed is made from fish meal. This, in turn, is made from fish which is often captured extra. This is in conflict with the sustainability goals of DRAPS. For this weakness already one opportunity is detected (see 10.1.3, i).

v. Personal responsibility

One issue in terms of operating large-scale aquaponics is the personnel responsibility. Currently, two options are discussed how DRAPS can be managed. The question is if it is better to operate DRAPS consolidated as one company or is a cooperation of two independent companies the better way? The community has widely differing opinions and further long-term studies would be necessary to give the right recommendation.

10.1.3 Opportunities

i. Market shift towards sustainability

Resource conservation is becoming increasingly important. Therefore one opportunity of DRAPS might be that there will be an increase appreciation for sustainable and locally produced food by consumers, especially in developed countries. This might result in higher sales prices when the environmental benefits are recognised by consumers or retailers. For example, it was noticed that consumers passing the DRAPS greenhouse in Waren (Germany; demonstration facility) were willing to pay high prices for tomatoes when they bought directly in the farm shop. The greenhouse was directly in the opposite of the farm shop. In this context, however, the marketing strategy would be highly important so that it can be valued by consumers and gain higher prices, especially by direct marketing.

Some scientist currently develops more sustainable fish feed, e.g. made from insects. Since mid of 2017 fish feed made from insects has been approved. This opportunity may solve the determined weakness of using not sustainable produced fish feed (see 10.1.2, *iv*).

ii. Sustainable intensification

Globally, the "sustainable intensification" is one approach to meet the century´s key challenge, food security. Due to the intensive, but sustainable food production, DRAPS can contribute as one instrument to sustainable intensification.

10.1.4 Threats

i. Consumers acceptance

One threat is that aquaponics is not known by many consumers. They do not know the idea behind aquaponics, the sustainable production system, and the benefits. On the other hand, consumers who know aquaponics, often ask if vegetables can be contaminated by fish faeces or bacteria when it is produced in aquaponic systems.

Another threat is that the consumers do not recognise or appreciate the sustainability of the system or that they even reject the system because the need of such systems is not realised.

ii. Old manners are the best proven

The question is: is the hydroponic world already ready to switch to aquaponics? The big companies producing vegetables in greenhouses have successfully established systems to gain high yields. A shift to aquaponics holds many dangers and the wisdom "old manners are the best proven" applies. However, an increased request for more sustainably produced products by the consumers (see 10.1.3), may change their minds.

In summary, even if competitiveness to conventional production systems in terms of quantity and quality could be proven, DRAPS seems to be not yet ready for use for commercial large-scale production. DRAPS are highly complex systems and easy technical solutions to optimise the fish waste water for the plants are missing. Additionally, for operating DRAPS successfully highly committed labour is required. Even if there a lack of know-how to operate DRAPS comprehensively at a level that is potentially possible (see above, weaknesses), the last years of research have led to knowledge gain about the complexity of DRAPS.

10.2 Summary and evaluation of improvements of double recirculating aquaponic systems

As stated in chapter 3, the DRAPS prototype presented by Kloas et al. (2015) was reconstructed and enhanced for the INAPRO project and therefore for the present studies. These enhancements included i) the implementation of a 3-cp and therewith the use of fish waste water gained by cleaning the mechanical filter (water saving aspects), ii) the spatial separation of fish and plant production, iii) a reduced fish to plant ratio, and iv) the continuous adjustment of the fish waste water used to prepare the nutrient solutions for the plants.

However, during the experimental period in the newly constructed research DRAPS, it became apparent that further modifications were necessary to improve the INAPRO DRAPS. As demonstrated in chapter 6 and 7 the installation of the 3-cp resulted in an undesired loss of nitrogen. This problem was solved by the implementation of a newly constructed suction filter device, described also in chapter 6 and 7. The new suction filter device enabled the direct water transfer from the RAS to the storage tank in the greenhouse. As a consequence, the 3-cp became redundant. Finally, it can be concluded that the implementation of the 3-cp was not covered with success but could be solved by the new suction device.

The other enhancements of the prototype (presented by Kloas et al. (2015)) can be indicated as improvements.

The spatial separation of the fish and the plants allowed growing the fish in a more suitable environment. While in the INAPRO research DRAPS the fish hall was also a working space for the staff, the fish hall was completely separated in the later built INAPRO demonstration DRAPS. Consequently, the environment for the fish was dark, quiet, and only with a few disturbances.

However, the reduced fish to plant ratio was generally insufficient. During the experiments only a part of the plants was irrigated with fish waste water (aquaponic treatments). But even if all plants in the greenhouse were irrigated, it would not have been enough to use all of the whole fish waste water delivered by RAS. However, for experimental use, the ratio was much better than in the prototype. The adjustment of the nutrient solution succeeded also much better than in the prototype. This was due to the regular analysis of the nutrient concentration in the process waters (tap water and fish waste water) and the frequent adjustment of the respective nutrient stock solution.

The implementation of the new suction device reduced not only the nitrogen loss but could also improve the quality of the nutrient solution in terms of the oxygen concentration. In the regarding experiment 1 (chapter 5) it was found that mainly the fish water transfer through the 3-cp increased the oxygen consumption of the fish water. This, in turn, could be the reason for the decreased oxygen concentration in the flowing nutrient solution through the cultivation trenches (subsection 5.4.1, Figure 11).

10.3 Outlook and recommendations for further research

Aquaponic research is a long way from its end. More or less all weaknesses and threats listed in chapter 10.1 are subjects which have to proceed further. Besides a selection from the mentioned points, a few other things will be outlined in this subsection.

10.3.1 Plant production in double recirculating aquaponic systems

It was found that some authors came to different results than those gained in the present thesis. Delaide et al. (2016) and Goddek and Vermeulen (2018) found a significantly increased yield of lettuce produced in DRAPS in comparison to conventional hydroponics. In the present thesis, the yields of tomatoes (chapter 3 and 6) and lettuce (chapter 8) was equal in DRAPS and conventional hydroponics. Further investigations should be done to ascertain why the results are different.

10.3.2 Connection between fish and plant

This thesis demonstrates that fish and plant production can be separately optimised and that DRAPS are a contribution to sustainable intensification of food production. However, a high potential for optimising DRAPS lies in nutrient management. As extensively discussed in chapter 9.2, further examinations are necessary to overcome the bottleneck of nutrient adjustment for the hydroponic unit. Therefore further studies on nutrient dynamics in RAS and research on robust and affordable sensors for an automatic nutrient adjustment is highly recommended. A smooth adjustment of the nutrient solution would be a very significant improvement of DRAPS. It would help to implement DRAPS as a commercial food production system.

In a first approach it was demonstrated that the nutrient concentration in the fish water fluctuates (subsection 9.2). The new suction device probably improved the stability of the nutrient concentrations. However, the investigation of the nutrient dynamics and the identification/understanding of influencing factors by fish rearing management is one essential topic for follow-up research.

Another issue is the strategy for use of fish waste water with already high nitrogen concentrations. As in subsection 9.2 extensive discussed, high nitrogen concentrations can be a problem for the addition of Ca and K fertiliser. Further research must be done to develop new strategies. One strategy could be the application of Ca by foliar fertilisation.

A further assumption was that the oxygen concentration in the cultivation trenches does not drop to low concentrations when the modified DRAPS is used (compare chapter 5). Oxygen depletion was associated with the 3-cp. Therefore it follows that leaving out the 3-cp could reduce the O_2 consumption and therewith improve the O_2-supply of the plants. Because this was just an assumption this should be investigated in further experiments.

10.3.3 Sustainability issues

DRAPS were developed to provide an answer for intensive but sustainable food production. Therefore maximal nutrient recycling and a zero waste approach should be pursued. In this context, nutrient remobilisation from sludge as well as a zero-discharge system in the hydroponic unit should be approached further. Because the nutrient remobilization from sludge was already adequately discussed by Goddek (2017) and Monsees (2018) it was not discussed further in the present study. However, the zero-discharge system is currently also an issue in conventional hydroponic plant production. Due to the accumulation of nutrients in the nutrient solution tanks a regular discharge of the "used/recirculated" nutrient solution is necessary. To meet the goal of minimal nutrient release in the environment and minimal water loss the necessary discharge should be minimised.

A further approach to increase the sustainability of DRAPS would be the use of more sustainable fish feed produced from insects (see 10.1.2 weaknesses, *iv* Fish feed). First experiments showed already that insect meal might be a possible protein source to be used in aquaculture. However, in regards to DRAPS, the effect of insect meal on the water quality must be investigated.

The research DRAPS used for the present studies was a light version (without recondensation of evpotranspirating water from the plants). Therefore, information about the fresh water demand of the full version of DRAPS is missing. It is highly recommended to use the demonstration facilities with an active cooling system to calculate the fresh water demand of DRAPS.

Another approach to save fresh water is the use of rainwater, already for fish rearing. It should be proven if it is possible to use rainwater for the different fish species.

10.3.4 Fish to plant ratio

The fish to plant ratios were calculated from the small research DRAPS. The ratio depends on various things which were discussed in subsection 9.3. More studies on this, using different fish and plant species, are necessary to describe the right ratio and the affecting factors.

10.3.5 Large-scaled systems

For the implementation of DRAPS as commercial large-scale production system, further research is necessary. Small experimental systems do not reflect the main constraints of large-scale systems. Furthermore, the production and investigations in large-scale systems would help to identify how DRAPS can be managed at its best and how economically they operate. Additionally, educational purposes can be derived from running commercial systems.

10.3.6 Intercropping

Further research is also necessary on intercropping or interplanting systems to bridge the plant production breaks. The intercropping of tomatoes with lettuce in the winter break is just one idea. Therefore, feasibility studies are necessary. In this context also the evaluation of the economic viability of intercropping is needed.

10.3.7 Consumers acceptance and fish welfare

DRAPS follows the sustainable approach. For success, it must be accepted by the consumers. Therefore, fish welfare is an issue which must also be considered. The consequent separation of fish and plants (in demonstration facilities) was a first step to increase fish welfare. But to get consumers´ acceptance the fish welfare in RAS should be exanimated e.g. under consideration of stocking density and nitrate concentration. The results of maximal fish welfare must be finally bought into context with DRAPS.

How do the stocking density and the maximal nitrate concentration affect the nutrient concentration in the fish waste water?

To increase the consumers' acceptance it is important to increase the awareness level of aquaponics, on the one hand. On the other hand consumers have to sensitise for environmental issues and sustainable food production systems like DRAPS. Another aspect for consumers´ acceptance is that the products, e.g. vegetables, are not contaminated by pathogenic microorganisms. Even if the aquaponics community does not assume problems in this context, the contamination issue is commonly mentioned/asked by consumers. Therefore more studies on this are necessary to provide further scientific evidence.

Acknowledgements

We gratefully acknowledge funding support from the European Commission via INAPRO (grant agreement number 619137).

11

References

11 References

Adams (1992). Crop nutrition in hydroponics. *Acta Horticulturae, 323*, 289–306.

Adams (1999). Plant nutrition demystified. *Acta Horticulturae, 481*, 341–344.

Adani, Genevini, Zaccheo, & Zocchi (1998). The effect of commercial humic acid on tomato plant growth and mineral nutrition. *Journal of plant nutrition, 21*(3), 561–575.

Adeniyi, Orjiekwe, Ehiagbonare, & Josiah (2012). Nutritional Composition of three different fishes (*Clarias gariepinus*, *Malapterurus electricus* and *Tilapia guineenisis*). *Pakistan Journal of Nutrition, 8*, 793–797.

Adeyeye (2009). Amino acid composition of three species of Nigerian fish: Clarias anguillaris, Oreochromis niloticus and Cynoglossus senegalensis. *Food Chemistry, 113*(1), 43–46.

Agarwal, & Rao (2000). Tomato lycopene and its role in human health and chronic diseases. *Canadian Medical Association Journal, 163*(6), 739–744.

Alam (1981). Effects of solution pH on the growth and chemical composition of rice plants. *Journal of Plant Nutrition, 4*(3), 247–260.

Almeida, Achten, Verbist, Heuts, Schrevens, & Muys (2014). Carbon and water footprints and energy use of greenhouse tomato production in Northern Italy. *Journal of Industrial Ecology, 18*(6), 898–908.

Almeselmani, Pant, & Singh (2009). Potassium level and physiological response and fruit quality in hydroponically grown tomato. *International Journal of Vegetable Science, 16*(1), 85–99.

Amann, Ludwig, & Schleifer (1995). Phylogenetic identification and in situ detection of individual microbial cells without cultivation. *Microbiological Reviews, 59*, 143–169.

AMI. (2017). *AMI Markt Bilanz Gemüse 2018.*

Anjana, & Iqbal (2007). Nitrate accumulation in plants, factors affecting the process, and human health implications. A review. *Agronomy for Sustainable Development, 27*(1), 45–57.

Antón, Montero, Muñoz, & Castells (2004). Identification of the main factors affecting the environmental impact of passive greenhouses. *Acta Horticulturae*, 489–494.

Antoniou, Hamilton, Koopman, Jain, Holloway, Lyberatos, & Svoronos (1990). Effect of temperature and pH on the effective maximum specific growth rate of nitrifying bacteria. *Water Research, 24*(1), 97–101.

Ast, & Draaijer. (2014). Methods and techniques to measure molecular oxygen in plants. In *Low-Oxygen Stress in Plants* (Vol. 21, pp. 397–417). Vienna: Springer.

ASU-L-26.11.03-4 (1983). Determination of the total acid content in tomato concentrate (potentiometric method).

Ávila-Juárez, Torres-Pacheco, Mercado-Luna, & Ocampo-Velázquez (2012). Tomato greenhouse productivity using interplanting system. *Acta Horticulturae, 947*, 133–138.

Badiola, Mendiola, & Bostock (2012). Recirculating Aquaculture Systems (RAS) analysis: Main issues on management and future challenges. *Aquacultural Engineering, 51*, 26–35.

Baldwin, Scott, Einstein, Malundo, Carr, Shewfelt, & Tandon (1998). Relationship between sensory and instrumental analysis for tomato flavor. *Journal of the American Society for Horticultural Science, 123*(5), 906–915.

Bamsey, Graham, Thompson, Berinstain, Scott, & Dixon (2012). Ion-specific nutrient management in closed systems: The necessity for ion-selective sensors in terrestrial and space-based agriculture and water management systems. *Sensors, 12*(10), 13349–13392.

Barbosa, Gadelha, Kublik, Proctor, Reichelm, Weissinger, . . . Halden (2015). Comparison of land, water, and energy requirements of lettuce grown using hydroponic vs. conventional agricultural methods. *International Journal of Environmental Research and Public Health, 12*(6), 6879–6891.

bbv. (2014). Erste Aquakultur-Anlage für tropischen Speisefisch in Bayern eröffnet. In B. Bauernverband (Ed.). www.proplanta.de.

Becker, Urlić, Špika, Kläring, Krumbein, Baldermann, . . . Schwarz (2015). Nitrogen limited red and green leaf lettuce accumulate flavonoid glycosides, caffeic acid derivatives, and sucrose while losing chlorophylls, β-carotene and xanthophylls. *PloS one, 10*(11), e0142867.

Bénard, Gautier, Bourgaud, Grasselly, Navez, Caris-Veyrat, . . . Genard (2009). Effects of low nitrogen supply on tomato (*Solanum lycopersicum*) fruit yield and quality with special emphasis on sugars, acids, ascorbate, carotenoids, and phenolic compounds. *Journal of Agricultural and Food Chemistry, 57*(10), 4112–4123.

Blancheton, Piedrahita, Eding, Lemarie, Bergheim, Fivelstad, & Roque D'Orbcastel (2007). Intensification of landbased aquaculture production in single pass and reuse systems. *Aquacultural Engineering and Environment*, 21–47.

Bloom, & Epstein (1984). Varietal differences in salt induced respiration in barley. *Plant science letters, 35*(1), 1–3.

Bostock, McAndrew, Richards, Jauncey, Telfer, Lorenzen, . . . Gatward (2010). Aquaculture: Global status and trends. *Philosophical Transactions of the Royal Society of London B: Biological Sciences, 365*(1554), 2897–2912.

Boudet (2007). Evolution and current status of research in phenolic compounds. *Phytochemistry, 68*(22), 2722–2735.

Boyd, McNevin, Clay, & Johnson (2005). Certification issues for some common aquaculture species. *Reviews in Fisheries Science, 13*(4), 231–279.

Bradford, & Hsiao (1982). Stomatal behavior and water relations of waterlogged tomato plants. *Plant Physiology, 70*(5), 1508–1513.

Brady, Hedlund, Cong, Hemerik, Hotes, Machado, . . . Thomsen (2015). Valuing supporting soil ecosystem services in agriculture: A natural capital approach. *Agronomy Journal, 107*(5), 1809–1821.

Bregnballe (2010). A guide to recirculation aquaculture. *An introduction to the new environmentally friendly and highly productive closed fish farming systems. Eurofish, Copenhagen, Denmark*.

Britto, & Kronzucker (2002). NH_4^+ toxicity in higher plants: A critical review. *Journal of Plant Physiology, 159*(6), 567–584.

Buchanan, & Omaye (2013). Comparative study of ascorbic acid and tocopherol concentrations in hydroponic-and soil-grown lettuces. *Food and Nutrition Sciences, 4*(10), 1047.

Carpenter (1966). New measurements of oxygen solubility in pure and natural water. *Limnology and Oceanography, 11*(2), 264–277.

Causse, Buret, Robini, & Verschave (2003). Inheritance of nutritional and sensory quality traits in fresh market tomato and relation to consumer preferences. *Journal of Food Science, 68*(7), 2342–2350.

Cecílio Filho, Rezende, Barbosa, & Grangeiro (2011). Agronomic efficiency of intercropping tomato and lettuce. *Anais da Academia Brasileira de Ciências, 83*(3), 1109–1119.

Cecílio Filho, Rezende, & Costa (2010). Economic analysis of the intercropping of lettuce and tomato in different seasons under protected cultivation. *Horticultura Brasileira, 28*(3), 326–336.

Chérif, Tirilly, & Bélanger (1997). Effect of oxygen concentration on plant growth, lipidperoxidation, and receptivity of tomato roots to *Pythium* F under hydroponic conditions. *European Journal of Plant Pathology, 103*(3), 255–264.

Chun, & Takakura (1994). Rate of root respiration of lettuce under various dissolved oxygen concentrations in hydroponics. *Environment Control in Biology, 32*(2), 125–135.

Claussen (2002). Growth, water use efficiency, and proline content of hydroponically grown tomato plants as affected by nitrogen source and nutrient concentration. *Plant and Soil, 247*(2), 199–209.

Coolong, & Randle (2003). Zinc concentration in hydroponic solution culture influences zinc and sulfur accumulation in *Brassica rapa* L. *Journal of plant nutrition, 26*(5), 949–959.

Correll (1998). The role of phosphorus in the eutrophication of receiving waters: A review. *Journal of Environmental Quality, 27*(2), 261–266.

Cripps, & Bergheim (2000). Solids management and removal for intensive land-based aquaculture production systems. *Aquacultural Engineering, 22*(1), 33–56.

d'Orbcastel, Blancheton, & Aubin (2009). Towards environmentally sustainable aquaculture: Comparison between two trout farming systems using life cycle assessment. *Aquacultural Engineering, 40*(3), 113–119.

Dannehl, Huyskens-Keil, Wendorf, Ulrichs, & Schmidt (2012). Influence of intermittent-direct-electric-current (IDC) on phytochemical compounds in garden cress during growth. *Food Chemistry, 131*(1), 239–246.

Dannehl, & Josuttis. (2014). Cultivar and production effects on bioactive polyphenols. In R. R. Watson (Ed.), *Polyphenols in Plants* (1 ed., pp. 3–13). USA: Academic Press

Dannehl, Josuttis, Huyskens-Keil, Ulrichs, & Schmidt (2014a). Comparison of different greenhouse systems and their impacts on plant responses of tomatoes. *Gesunde Pflanzen, 66*(3), 111–119.

Dannehl, Josuttis, Ulrichs, & Schmidt (2014b). The potential of a confined closed greenhouse in terms of sustainable production, crop growth, yield and valuable plant compounds of tomatoes. *Journal of Applied Botany and Food Quality, 87*.

Dannehl, Rocksch, & Schmidt (2015a). Modelling to estimate the specific leaf area of tomato leaves ('Pannovy'). *Acta Horticulturae, 1099*, 79–86.

Dannehl, Schuch, & Schmidt (2013). Plant production in solar collector greenhouses – influence on yield, energy use efficiency and reduction in CO_2 emissions. *Journal of Agricultural Science, 5*(10), 34–45.

Dannehl, Suhl, Huyskens-Keil, Ulrichs, & Schmidt (2014c). Effects of a special solar collector greenhouse on water balance, fruit quantity and fruit quality of tomatoes. *Agricultural Water Management, 134*, 14–23.

Dannehl, Suhl, Ulrichs, & Schmidt (2015b). Evaluation of substitutes for rock wool as growing substrate for hydroponic tomato production. *Journal of Applied Botany and Food Quality, 88*, 68–77.

Datko, Mudd, Giovanelli, & Macnicol (1978). Sulfur-containing compounds in *Lemna perpusilla* 6746 grown at a range of sulfate concentrations. *Plant physiology, 62*(4), 629–635.

Davies, Lloyd, & Boddy (1989). The effect of oxygen on denitrification in *Paracoccus denitrificans* and *Pseudomonas aeruginosa*. *Microbiology, 135*(9), 2445–2451.

De Kreij, Voogt, van den Bos, & Baas. (1999). *Bemestings adviesbasis substraten*: PBG Naaldwijk.

Delaide. (2017). *A study on the mineral elements available in aquaponics, their impact on lettuce productivity and the potential improvement of their availability.* (Doctoral Thesis), Université de Liège, Liège.

Delaide, Goddek, Gott, Soyeurt, & Jijakli (2016). Lettuce (*Lactuca sativa* L. var. Sucrine) growth performance in complemented aquaponic solution outperforms hydroponics. *Water, 8*(10), 467.

Demsar, & Osvald (2003). Influence of NO_3^-: NH_4^+ ratio on growth and nitrate accumulation in lettuce (*Lactuca sativa* var. capitata L.) in an aeroponic system. *Agrochimica, 47*(3–4), 112–121.

Di Mascio, Kaiser, & Sies (1989). Lycopene as the most efficient biological carotenoid singlet oxygen quencher. *Archives of Biochemistry and Biophysics, 274*(2), 532–538.

DIN-ISO-10694 (1995). Soil quality – Determination of organic and total carbon after dry combustion (elementary analysis). *ISO 10694:1995.*

DIN-ISO-13878 (1998). Soil quality – Determination of total nitrogen content by dry combustion (elemental analysis). *ISO 13878:1998.*

Diver, & Rinehart (2010). Aquaponics – integration of hydroponics with aquaculture.

Dorai, Papadopoulos, & Gosselin (2001). Influence of electric conductivity management on greenhouse tomato yield and fruit quality. *Agronomie, 21*(4), 367–383.

Dumas, Dadomo, Di Lucca, & Grolier (2003). Effects of environmental factors and agricultural techniques on antioxidantcontent of tomatoes. *Journal of the Science of Food and Agriculture, 83*(5), 369–382.

DuPont, Mondin, Williamson, & Price (2000). Effect of variety, processing, and storage on the flavonoid glycoside content and composition of lettuce and endive. *Journal of agricultural and food chemistry, 48*(9), 3957–3964.

Eck, Sare, Massart, Schmautz, Junge, Smits, & Jijakli (2019). Exploring bacterial communities in aquaponic systems. *Water, 11*(2), 260.

Economakis (1993). The influence of solution heating and intermittent solution circulation on tomatoes in nutrient film culture. *Acta Horticulturae, 323*, 81–88.

Economakis, & Koleilat. (1996). *Effect of nitrogen concentration on growth, water, and nutrient uptake of lettuce plants in solution culture.* Paper presented at the II International Symposium on Irrigation of Horticultural Crops 449.

Edenhofer, Pichs-Madruga, Sokona, Farahani, Kadner, Seyboth, . . . Minx (2014). Climate Change 2014: Mitigation of climate change. *Contribution of Working Group III to the Fifth Assessment. Report of the Intergovernmental Panel on Climate Change*.

Ehret, Alsanius, Wohanka, Menzies, & Utkhede (2001). Disinfestation of recirculating nutrient solutions in greenhouse horticulture. *Agronomie, 21*(4), 323–339.

Endut, Lananan, Abdul Hamid, Jusoh, & Wan Nik (2016). Balancing of nutrient uptake by water spinach (*Ipomoea aquatica*) and mustard green (*Brassica juncea*) with nutrient production by African catfish (*Clarias gariepinus*) in scaling aquaponic recirculation system. *Desalination and Water Treatment, 57*(60), 29531–29540.

Engels, Kirkby, & White. (2012). Mineral nutrition, yield and source-sink relationships. In P. Marschner (Ed.), *Marschner´s Mineral Nutrition of Higher Plants* (Vol. 3, pp. 85–133): Elsevir Ltd.

Evans, Piermarini, & Choe (2005). The multifunctional fish gill: Dominant site of gas exchange, osmoregulation, acid-base regulation, and excretion of nitrogenous waste. *Physiological Reviews, 85*(1), 97–177.

Fageria (2001). Nutrient interactions in crop plants. *Journal of plant nutrition, 24*(8), 1269–1290.

Falk, Tokuhisa, & Gershenzon (2007). The effect of sulfur nutrition on plant glucosinolate content: Physiology and molecular mechanisms. *Plant Biology, 9*(05), 573–581.

FAO. (2009). *Global agriculture towards 2050*. Retrieved from http://www.fao.org/fileadmin/templates/wsfs/docs/Issues_papers/HLEF2050_Global_Agriculture.pdf

FAO. (2010). *Cultured aquatic species information programme. Clarias gariepinus*. Retrieved from Rome: http://www.fao.org/fishery/culturedspecies/Clarias_gariepinus/en

FAO. (2011). *The state of the world's land and water resources for food and agriculture – managing systems at risk*. Retrieved from US & Canada: www.fao.org/3/a-i1688e.pdf

FAO. (2014). *The state of world fisheries and aquaculture – opportunities and challanges*. Retrieved from Rome: www.fao.org/3/a-i3720e.pdf

FAO. (2016a). *AQUASTAT database.* Retrieved from http://www.fao.org/nr/water/aquastat/tables/WorldData-Withdrawal_eng.pdf

FAO. (2016b). *The state of world fisheries and aquaculture – contributing to food security and nutrition for all*. Retrieved from Rome: www.fao.org/3/a-i5555e.pdf

Fish, Perkins-Veazie, & Collins (2002). A quantitative assay for lycopene that utilizes reduced volumes of organic solvents. *Journal of Food Composition and Analysis, 15*(3), 309–317.

Förster, Ulrichs, Schreiner, Arndt, Schmidt, & Mewis (2015). Ecotype variability in growth and secondary metabolite profile in *Moringa oleifera*: Impact of sulfur and water availability. *Journal of agricultural and food chemistry, 63*(11), 2852–2861.

Frusciante, Carli, Ercolano, Pernice, Di Matteo, Fogliano, & Pellegrini (2007). Antioxidant nutritional quality of tomato. *Molecular Nutrition & Food Research, 51*(5), 609–617.

Ghaly, Kamal, & Mahmoud (2005). Phytoremediation of aquaculture wastewater for water recycling and production of fish feed. *Environment International, 31*(1), 1–13.

Giovannucci (1999). Tomatoes, tomato-based products, lycopene, and cancer: Review of the epidemiologic literature. *Journal of the National Cancer Institute, 91*(4), 317–331.

Gislerød, & Adams (1983). Diurnal variations in the oxygen content and acid requirement of recirculating nutrient solutions and in the uptake of water and potassium by cucumber and tomato plants. *Scientia Horticulturae, 21*(4), 311–321.

Gislerød, & Kempton (1983). The oxygen content of flowing nutrient solutions used for cucumber and tomato culture. *Scientia Horticulturae, 20*(1), 23–33.

Gitelson, & Merzlyak (1994). Spectral reflectance changes associated with autumn senescence of *Aesculus hippocastanum* L. and *Acer platanoides* L. leaves. Spectral features and relation to chlorophyll estimation. *Journal of Plant Physiology, 143*(3), 286–292.

Goddek. (2017). *Opportunities and challenges of multi-loop aquaponic systems.* (Doctoral thesis), Wageningen University, Wageningen, the Netherlands.

Goddek, Delaide, Mankasingh, Ragnarsdottir, Jijakli, & Thorarinsdottir (2015). Challenges of sustainable and commercial aquaponics. *Sustainability, 7*(4), 4199–4224.

Goddek, Espinal, Delaide, Jijakli, Schmautz, Wuertz, & Keesman (2016a). Navigating towards decoupled aquaponic systems: A system dynamics design approach. *Water, 8*(7), 303.

Goddek, Schmautz, Scott, Delaide, Keesman, Wuertz, & Junge (2016b). The effect of anaerobic and aerobic fish sludge supernatant on hydroponic lettuce. *Agronomy, 6*(2), 37.

Goddek, & Vermeulen (2018). Comparison of *Lactuca sativa* growth performance in conventional and RAS-based hydroponic systems. *Aquaculture International*, 1–10.

Godfray, Beddington, Crute, Haddad, Lawrence, Muir, . . . Toulmin (2010). Food security: The challenge of feeding 9 billion people. *Science, 327*(5967), 812–818.

Goto, Both, Albright, Langhans, & Leed (1996). Effect of dissolved oxygen concentration on lettuce growth in floating hydroponics. *Acta Horticulturae, 440*, 205–210.

Graber, & Junge (2009). Aquaponic systems: Nutrient recycling from fish wastewater by vegetable production. *Desalination, 246*(1), 147–156.

Grafiadellis, Mattas, Maloupa, Tzouramani, & Galanopoulos (2000). An economic analysis of soilless culture in gerbera production. *HortScience, 35*(2), 300–303.

Gunes, Aktas, & Post (1995). Effect of partial replacement of nitrate by NH_4-N, urea-N and amino acid-N in nutrient solution on nitrate accumulation in lettuce (*Lactuca sativa* L.). *Agrochimica (Italy)*.

Gutzmann, & Kledal. (2014). *Aquaponics and the prospects of large scale commercial production*. Paper presented at the European Aquaculture Society â Adding Value, Donastia â San Sebastian, Spain.

Hambly, Arvin, Pedersen, Pedersen, Seredyńska-Sobecka, & Stedmon (2015). Characterising organic matter in recirculating aquaculture systems with fluorescence EEM spectroscopy. *Water research, 83*, 112–120.

Hawkesford, Horst, Kichey, Lambers, Schjoerring, Skrumsager Møller, & White. (2012). Functions of macronutrients. In P. Marschner (Ed.), *Marchner´s mineral nutrition of higher plants* (Vol. 3, pp. 135–189): Elsevier Ltd.

Herrmann (1979). Übersicht über die Inhaltsstoffe der Tomaten. *Zeitschrift für Lebensmittel-Untersuchung und Forschung, 169*(3), 179–200.

Hickman. (2018). International greenhouse vegetable production – Statistics (2018 Edition). Retrieved 07.09.2018, from Cuesta Roble Greenhouse Vegetable Consulting cuestaroble.com/statistics.htm

Hobiger (1996). Ammoniak im Wasser – Anleitung einer Formel zur Berechnung von Ammoniak in wäßrigen Lösungen. *Umweltbundesamt UBA-BE-076*.

Hochheimer, & Wheaton. (1998). *Biological filters: Trickling and RBC design.* Paper presented at the Proceedings of the Second International Conference on Recirculating Aquaculture, Roanoke, VA.

Hochmuth. (2001). *Fertilizer management for greenhouse vegetables* (U. o. F. I. Extension Ed. Vol. 3). Gainesville, Florida: University of Florida.

Hochmuth, Maynard, Vavrina, Hanlon, & Simonne. (2004). Plant tissue analysis and interpratation for vegetable crops in Florida. In: University of Florida.

Hoffland, Dicke, Van Tintelen, Dijkman, & Van Beusichem (2000a). Nitrogen availability and defense of tomato against two-spotted spider mite. *Journal of Chemical Ecology, 26*(12), 2697–2711.

Hoffland, Jeger, & van Beusichem (2000b). Effect of nitrogen supply rate on disease resistance in tomato depends on the pathogen. *Plant and Soil, 218*(1), 239–247.

Holtman, Oppedijk, & Draaijer (2009a). Development and demonstration of an optical oxygen sensor for horticulture. *Acta Horticulturae, 843*, 43–48.

Holtman, Oppedijk, Vennik, & Draaijer (2009b). Development of an oxygen measurement system as a management tool in horticulture. *Acta Horticulturae, 819*, 257–264.

Honda, Watanabe, Kikuchi, Iwata, Takeda, Uemoto, . . . Kiyono (1993). High density rearing of Japanese flounder, *Paralichthys olivaceus* with a closed seawater recirculation system equipped with a denitrification unit. *Aquaculture Science, 41*(1), 19–26.

hortidaily.com. (2015, 09.04.2018). Belgian tomato grower delivers heat and electricity to fish farm. *hortidaily.com*.

Howarth, & Marino (2006). Nitrogen as the limiting nutrient for eutrophication in coastal marine ecosystems: Evolving views over three decades. *Limnology and Oceanography, 51*(1part2), 364–376.

Ip, Chew, & Randall (2001). Ammonia toxicity, tolerance, and excretion. *Fish physiology, 20*, 109–148.

Islam, Edwards, & Asher (1980). pH optima for crop growth. *Plant and Soil, 54*(3), 339–357.

Jackson (1980). Aeration in the nutrient film technique of glasshouse crop production and the importance of oxygen, ethylene and carbon dioxide. *Acta Horticulturae, 98*, 61–78.

Jauset, Sarasúa, Avilla, & Albajes (2000). Effect of nitrogen fertilization level applied to tomato on the greenhouse whitefly. *Crop Protection, 19*(4), 255–261.

Jensen (1999). Hydroponics worldwide. *Acta Horticulturae, 481*, 719–729.

Jenssen, & Kongshaug (2003). Energy consumption and greenhouse gas emissions in fertilizer production. *International Fertiliser Society, 509.*

Jett, Chism, & Conley. (2005). Intercropping systems for tomatoes within a high tunnel. Retrieved 04.08.2017 https://www.researchgate.net/profile/Shawn_Conley/publication/266182369_Intercropping_Systems_for_Tomatoes_within_a_High_Tunnel/links/54d21f7b0cf28370d0e1bd92/Intercropping-Systems-for-Tomatoes-within-a-High-Tunnel.pdf

Jung, Kim, Choi, Ahn, Son, & Sudduth (2015). Automated lettuce nutrient solution management using an array of ion-selective electrodes. *Transactions of the ASABE, 58*(5), 1309–1319.

Junge, König, Villarroel, Komives, & Jijakli (2017). Strategic points in aquaponics. *Water, 9*(3), 182.

Karimanzira, Keesman, Kloas, Baganz, & Rauschenbach (2016a). Dynamic modeling of the INAPRO aquaponic system. *Aquacultural Engineering, 75*(Supplement C), 29–45.

Karimanzira, Keesman, Kloas, Baganz, & Rauschenbach (2016b). Efficient and economical way of operating a recirculation aquaculture system in an aquaponics farm. *Aquaculture Economics & Management*, 1–18.

Kläring (2001). Strategies to control water and nutrient supplies to greenhouse crops. A review. *Agronomie, 21*(4), 311–321.

Kledal, & Thorarinsdottir. (2018). Aquaponics: A commercial niche for sustainable modern aquaculture. In *Sustainable Aquaculture* (pp. 173–190): Springer.

Klinger, & Naylor (2012). Searching for solutions in aquaculture: Charting a sustainable course. *Annual Review of Environment and Resources, 37*, 247–276.

Kloas, Groß, Baganz, Graupner, Monsees, Schmidt, . . . Rennert (2015). A new concept for aquaponic systems to improve sustainability, increase productivity, and to reduce environmental impacts. *Aquaculture Environment Interactions, 7*, 179–192.

Konstantopoulou, Kapotis, Salachas, Petropoulos, Karapanos, & Passam (2010). Nutritional quality of greenhouse lettuce at harvest and after storage in relation to N application and cultivation season. *Scientia Horticulturae, 125*(2), 93. e91–93. e95.

Krüner, & Rosenthal (1983). Efficiency of nitrification in trickling filters using different substrates. *Aquacultural Engineering, 2*(1), 49–67.

Kumar, & Cho (2014). Reuse of hydroponic waste solution. *Environmental Science and Pollution Research, 21*(16), 9569–9577.

Lähne. (2016). Deutschlands größtes Gewächshaus in Wittenberg soll vergrößert werden. *Leipziger Volkszeitung*.

Lattauschke. (2000). *Ökonomische Betrachtungen zum Anbau von Gemüse im Gewächshaus*. Retrieved from Dresden: https://publikationen.sachsen.de/bdb/artikel/13886

Lattauschke. (2004a). *Gewächshaustomaten – Hinweis zum umweltgerechten Anbau und Managementunterlagen*. Retrieved from Dresden, Germany:

Lattauschke. (2004b). *Salate im Gewächshaus – Hinweis zum umweltgerechten Anbau und Managementunterlage*. Retrieved from Dresden, Germany:

Lawson. (1994). *Fundamentals of aquacultural engeneering*. Boston/Dordrecht/London Kluwer Academic Publisher.

Le Bot, Jeannequin, & Fabre (2001). Impacts of N-deprivation on the yield and nitrogen budget of rockwool grown tomatoes. *Agronomie, 21*(4), 341–350.

Lehmann, Scheibe, & Jobs. (2015). Germany Patent No. DE 102014004767 A1. Pal-Anlagenbau GmbH Abtshagen.

Lekang, & Kleppe (2000). Efficiency of nitrification in trickling filters using different filter media. *Aquacultural Engineering, 21*(3), 181–199.

Leonard, Blancheton, & Guiraud (2000). Populations of heterotrophic bacteria in an experimental recirculating aquaculture system. *Aquacultural Engineering, 22*(1), 109–120.

Leonard, Guiraud, Gasset, Cailleres, & Blancheton (2002). Bacteria and nutrients—nitrogen and carbon—in a recirculating system for sea bass production. *Aquacultural Engineering, 26*(2), 111–127.

Licamele. (2009). *Biomass production and nutrient dynamics in an aquaponics system.* The University of Arizona,

Lindell, Falls, & Goldman (1996). Effluent treatment and residuals management using land application to recycle nutrients from a large-scale recirculating aquaculture facility. *Successes and Failures in Commercial Recirculating Aquaculture*, 240.

Llorach, Martínez-Sánchez, Tomás-Barberán, Gil, & Ferreres (2008). Characterisation of polyphenols and antioxidant properties of five lettuce varieties and escarole. *Food Chemistry, 108*(3), 1028–1038.

Love, Fry, Genello, Hill, Frederick, Li, & Semmens (2014). An international survey of aquaponics practitioners. *PloS one, 9*(7), e102662.

Lundberg, Weitzberg, Cole, & Benjamin (2004). Nitrate, bacteria and human health. *Nature Reviews Microbiology, 2*(7), 593–602.

MacArthur (1916). Solubility of oxygen in salt solutions and the hydrates of these salts. *The Journal of Physical Chemistry, 20*(6), 495–502.

MacCarthy (2001). The principles of humic substances. *Soil Science, 166*(11), 738–751.

Mackowiak, Stutte, Wheeler, Ruffe, & Yorio (1999). Tomato and soybean production on a shared recirculating hydroponic system. *Acta Horticulturae, 481*, 259–266.

Madigan, Martinko, & Parker. (2017). *Brock biology of microorganisms* (Vol. 13): Pearson.

Mageau, Radtke, Fazendin, & Ledin (2015). The aquaponics solution. *The Solutions Journal, 6*(3), 51–59.

Martins, Eding, Verdegem, Heinsbroek, Schneider, Blancheton, . . . Verreth (2010). New developments in recirculating aquaculture systems in Europe: A perspective on environmental sustainability. *Aquacultural Engineering, 43*(3), 83–93.

Martins, Pistrin, Ende, Eding, & Verreth (2009). The accumulation of substances in recirculating aquaculture systems (RAS) affects embryonic and larval development in common carp *Cyprinus carpio*. *Aquaculture, 291*(1), 65–73.

Masser, Rakocy, & Losordo (1992). Recirculating aquaculture tank production systems. Management of recirculating systems. *Southern Regional Aquaculture Center (SRAC) Publication, 452*.

Mattson, & Peters (2014). A recipe for hydroponic success. *Inside Grower. Ball Publishing, Chicago, IL*, 16–19.

Maucieri, Nicoletto, Junge, Schmautz, Sambo, & Borin (2017). Hydroponic systems and water management in aquaponics: A review. *Italian Journal of Agronomy, 11*.

McCall, & Willumsen (1998). Effects of nitrate, ammonium and chloride application on the yield and nitrate content of soil-grown lettuce. *The Journal of Horticultural Science and Biotechnology, 73*(5), 698–703.

McMurtry, Sanders, Patterson, & Nash (1993). Yield of tomato irrigated with recirculating aquacultural water. *Journal of Production Agriculture, 6*(3), 428–432.

Merino, Darnerud, Edberg, Åman, & Castillo (2006). Levels of nitrate in Swedish lettuce and spinach over the past 10 years. *Food Additives and Contaminants, 23*(12), 1283–1289.

Millikan, Bjarnason, Osborn, & Hanger (1971). Calcium concentration in tomato fruits in relation to the incidence of blossom-end rot. *Animal Production Science, 11*(52), 570–575.

Monsees. (2018). *Overcoming major bottlenecks in aquaponics – a practical approach.* (Doctoral thesis), Humboldt-Universität zu Berlin, Berlin, Germany. Retrieved from https://edoc.hu-berlin.de/handle/18452/19368

Monsees, Keitel, Paul, Kloas, & Wuertz (2017a). Potential of aquacultural sludge treatment for aquaponics: evaluation of nutrient mobilization under aerobic and anaerobic conditions. *Aquaculture Environment Interactions, 9*, 9–18.

Monsees, Klatt, Kloas, & Wuertz (2016). Chronic exposure to nitrate significantly reduces growth and affects the health status of juvenile Nile tilapia (*Oreochromis niloticus* L.) in recirculating aquaculture systems. *Aquaculture Research, 48*(7), 3482–3492.

Monsees, Kloas, & Wuertz (2017b). Decoupled systems on trial: Eliminating bottlenecks to improve aquaponic processes. *PloS One, 12*(9), e0183056.

Montagu, & Goh (1990). Effects of forms and rates of organic and inorganic nitrogen fertilisers on the yield and some quality indices of tomatoes (*Lycopersicon esculentum* Miller). *New Zealand Journal of Crop and Horticultural Science, 18*(1), 31–37.

Morard, Lacoste, & Silvestre (2000). Effect of oxygen deficiency on uptake of water and mineral nutrients by tomato plants in soilless culture. *Journal of Plant Nutrition, 23*(8), 1063–1078.

Morard, & Silvestre (1996). Plant injury due to oxygen deficiency in the root environment of soilless culture: A review. *Plant and Soil, 184*(2), 243–254.

Msangi, Kobayashi, Batka, Vannuccini, Dey, & Anderson. (2013). *Fish to 2030: Prospects for fisheries and aquaculture*. Retrieved from Washington DC: www.fao.org/3/i3640e/i3640e.pdf

Mulabagal, Ngouajio, Nair, Zhang, Gottumukkala, & Nair (2010). In vitro evaluation of red and green lettuce (*Lactuca sativa*) for functional food properties. *Food chemistry, 118*(2), 300–306.

Müritz-Wasser-/Abwasserzweckverband. (2011). *Gebührensatz Abwasser*. Retrieved from Waren (Müritz), Germany:

Murray, Bostock, & Fletcher. (2014). *Review of recirculating aquaculture system technologies and thier commercail application*. Retrieved from http://www.hie.co.uk/common/handlers/download-document.ashx?id=236008c4-f52a-48d9-9084-54e89e965573

Nagata, & Yamashita (1992). Simple method for simultaneous determination of chlorophyll and carotenoids in tomato fruit. *Journal of the Japanese Society for Food Science and Technology, 39*(10), 925–928.

Nazaryuk, Klenova, & Kalimullina (2002). Ecoagrochemical approaches to the problem of nitrate pollution in agroecosystems. *Russian Journal of Ecology, 33*(6), 392–397.

Nichols, & Savidov. (2011). *Aquaponics: A nutrient and water efficient production system.* Paper presented at the II International Symposium on Soilless Culture and Hydroponics 947.

Nielsen, Bouma, Lynch, & Eissenstat (1998). Effects of phosphorus availability and vesicular–arbuscular mycorrhizas on the carbon budget of common bean (*Phaseolus vulgaris*). *New Phytologist, 139*(4), 647–656.

Ostermeyer (1999). Vitamine in Fischen. *Informationen für die Fischwirtschaft aus der Fischereiforschung, 46*(3), 42–50.

Palm, Bissa, & Knaus (2014a). Significant factors affecting the economic sustainability of closed aquaponic systems. Part II: Fish and plant growth. *AACL Bioflux, 7*(3), 162–175.

Palm, Seidemann, Wehofsky, & Knaus (2014b). Significant factors affecting the economic sustainability of closed aquaponic systems. Part I: System design, chemo-physical parameters and general aspects. *AACL Bioflux, 7*(1), 20–32.

Palta, & Nobel (1989). Root respiration for Agave deserti: Influence of temperature, water status and root age on daily patterns. *Journal of Experimental Botany, 40*(2), 181–186.

Pandey, & Rizvi (2009). Plant polyphenols as dietary antioxidants in human health and disease. *Oxidative Medicine and Cellular Longevity, 2*(5), 270–278.

Pantanella, Cardarelli, Colla, Rea, & Marcucci (2012). Aquaponics vs. hydroponics: Production and quality of lettuce crop. *Acta Horticulturae, 927*, 887–893.

Pantanella, Cardarelli, Di Mattia, & Colla (2015). Aquaponics and food safety: Effects of UV sterilization on total coliforms and lettuce production. *Acta Horticulturae, 1062*, 71–76.

Pardossi, Tognoni, & Incrocci (2004). Mediterranean greenhouse technology. *Chronica Horticulturae, 44*(2), 28–34.

Patureau, Bernet, & Moletta (1996). Effect of oxygen on denitrification in continuous chemostat culture with *Comamonas* sp $SGLY_2$. *Journal of Industrial Microbiology, 16*(2), 124–128.

Patureau, Zumstein, Delgenes, & Moletta (2000). Aerobic denitrifiers isolated from diverse natural and managed ecosystems. *Microbial Ecology, 39*(2), 145–152.

Peet, & Welles. (2005). Greenhouse tomato production. In E. Heuvelink (Ed.), *Tomatoes* (Vol. 13, pp. 257–304): CABI Publishing.

Pollard, Kirk, & Cade (2002). Factors affecting food choice in relation to fruit and vegetable intake: A review. *Nutrition Research Reviews, 15*(02), 373–387.

Radi, Mahrouz, Jaouad, & Amiot (2003). Influence of mineral fertilization (NPK) on the quality of apricot fruit (cv. Canino). The effect of the mode of nitrogen supply. *Agronomie, 23*(8), 737–745.

Rakocy, Bailey, Shultz, & Thoman. (2004a). *Update on Tilapia and vegetable production in the UVI aquaponic system.* Paper presented at the New Dimensions on farmed Tilapia: Proceedings of the sixth international symposium on Tilapia in aquaculture, Held September.

Rakocy, Masser, & Losordo. (2006). *Recirculating aquaculture tank production systems: Aquaponics-integrating fish and plant culture.* (454). SRAC, Stoneville, Mississippi.

Rakocy, Shultz, Bailey, & Thoman. (2004b). *Aquaponic production of tilapia and basil: comparing a batch and staggered cropping system.* Paper presented at the South Pacific Soilless Culture Conference-SPSCC 648.

Rennert, Groß, van Ballegooy, & Kloas (2011). Ein Aquaponiksystem zur kombinierten Tilapia- und Tomatenproduktion. *Fischer & Teichwirt, 06*, 209–214.

Resh. (2004). *Hydroponic food production* (Vol. 6). Mahwah, New Jersey: Newconcept Press, Inc.

Reyes Lastiri, Slinkert, Cappon, Baganz, Staaks, & Keesman (2016). Model of an aquaponic system for minimised water, energy and nitrogen requirements. *Water Science and Technology, 74*(1), 30–37.

Ribas-Agustí, Gratacós-Cubarsí, Sárraga, García-Regueiro, & Castellari (2011). Analysis of eleven phenolic compounds including novel p-coumaroyl derivatives in lettuce (*Lactuca sativa* L.) by ultra-high-performance liquid chromatography with photodiode array and mass spectrometry detection. *Phytochemical Analysis, 22*(6), 555–563.

Richardson, Duigan, & Berlyn (2002). An evaluation of noninvasive methods to estimate foliar chlorophyll content. *New Phytologist, 153*(1), 185–194.

Roosta (2014). Comparison of the vegetative growth, eco-physiological characteristics and mineral nutrient content of basil plants in different irrigation ratios of hydroponic:aquaponic solutions. *Journal of Plant Nutrition, 37*(11), 1782–1803.

Roosta, & Ghorbani (2011). Investigation of the growth and development, essential oil and minerals content in two species of mint in hydroponics and aquaponics. *HortTechnology, 2*(7), 19–28.

Roosta, & Hamidpour (2011). Effects of foliar application of some macro- and micro-nutrients on tomato plants in aquaponic and hydroponic systems. *Scientia Horticulturae, 129*(3), 396–402.

Roosta, & Hamidpour (2013). Mineral nutrient content of tomato plants in aquaponic and hydroponic systems: Effect of foliar application of some macro-and micro-nutrients. *Journal of Plant Nutrition, 36*(13), 2070–2083.

Rouphael, Colla, Battistelli, Moscatello, Proietti, & Rea (2004). Yield, water requirement, nutrient uptake and fruit quality of zucchini squash grown in soil and closed soilless culture. *The Journal of Horticultural Science and Biotechnology, 79*(3), 423–430.

Ruijs, & Van Os (1990). Economic evaluation of business systems with a lower degree of environmental pollution. *Horticultural Economics and Marketing, XXIII IHC 295*, 79–84.

Salam, Asadujjaman, & Rahman (2013). Aquaponics for improving high density fish pond water quality through raft and rack vegetable production. *World Journal of Fish and Marine Sciences, 5*(3), 251–256.

Savidov, Hutchings, & Rakocy. (2005). *Fish and plant production in a recirculating aquaponic system: A new approach to sustainable agriculture in Canada*. Paper presented at the International Conference and Exhibition on Soilless Culture: ICESC 742.

Savvas (2003). Hydroponics: A modern technology supporting the application of integrated crop management in greenhouse. *Journal of Food Agriculture and Environment, 1*, 80–86.

Savvas, Stamati, Tsirogiannis, Mantzos, Barouchas, Katsoulas, & Kittas (2007). Interactions between salinity and irrigation frequency in greenhouse pepper grown in closed-cycle hydroponic systems. *Agricultural Water Management, 91*(1), 102–111.

Scalbert, & Williamson (2000). Dietary intake and bioavailability of polyphenols. *The Journal of Nutrition, 130*(8), 2073S-2085S.

Schiavon, Pizzeghello, Muscolo, Vaccaro, Francioso, & Nardi (2010). High molecular size humic substances enhance phenylpropanoid metabolism in maize (*Zea mays* L.). *Journal of chemical ecology, 36*(6), 662–669.

Schmautz, Graber, Jaenicke, Goesmann, Junge, & Smits (2017). Microbial diversity in different compartments of an aquaponics system. *Archives of Microbiology, 199*(4), 613–620.

Schmautz, Loeu, Liebisch, Graber, Mathis, Griessler Bulc, & Junge (2016). Tomato productivity and quality in aquaponics: Comparison of three hydroponic methods. *Water, 8*(11), 533.

Schram, Roques, Abbink, Spanings, de Vries, Bierman, . . . Flik (2010). The impact of elevated water ammonia concentration on physiology, growth and feed intake of African catfish (*Clarias gariepinus*). *Aquaculture, 306*(1), 108–115.

Schram, Roques, Abbink, Yokohama, Spanings, Vries, . . . Flik (2014). The impact of elevated water nitrate concentration on physiology, growth and feed intake of African catfish *Clarias gariepinus* (Burchell 1822). *Aquaculture Research, 45*(9), 1499–1511.

Schreier, Mirzoyan, & Saito (2010). Microbial diversity of biological filters in recirculating aquaculture systems. *Current Opinion in Biotechnology, 21*(3), 318–325.

Schröder, & Lieth. (2002). Irrigation control in hydroponics. In (pp. 263–298): Embryo Publications: Athens, Greece.

Siddiqi, Malhotra, Min, & Glass (2002). Effects of ammonium and inorganic carbon enrichment on growth and yield of a hydroponic tomato crop. *Journal of Plant Nutrition and Soil Science, 165*(2), 191.

Snow, Ghaly, & Snow (2008). A comparative assessment of hydroponieally grown cereal crops for the purification of aquaculture wastewater and the production of fish feed. *American Journal of Agricultural and Biological Sciences, 3*(1), 364–378.

Snyder. (2016). *Greenhouse tomato handbook* (M. S. University Ed.): Mississippi State University Extension Service.

Soffer, Burger, & Lieth (1991). Plant growth and development of *Chrysanthemum* and *Ficus* in aero-hydroponics: Response to low dissolved oxygen concentrations. *Scientia Horticulturae, 45*(3–4), 287–294.

Somerville, Cohen, Pantanella, Stankus, & Lovatelli. (2014). *Small-scale aquaponic food production – Integrated fish and plant farming*. Retrieved from Rome:

Sonneveld, & Welles (1988). Yield and quality of rockwool-grown tomatoes as affected by variations in EC-value and climatic conditions. *Plant and Soil, 111*(1), 37–42.

Stahl, & Sies (1996). Lycopene: A biologically important carotenoid for humans? *Archives of Biochemistry and biophysics, 336*(1), 1–9.

Stanghellini, & Kempkes. (2004). *A blueprint for optimal management of multiple-quality water-resources*. Retrieved from https://www.aua.gr/ns/project/hortimed/Deliverable_8.pdf

Stanghellini, Kempkes, & Knies (2003). Enhancing environmental quality in agricultural systems. *Acta Horticulturae, 609*, 277–283.

Steffens, & Wirth (2005). Freshwater fish-an important source of n-3 polyunsaturated fatty acids: A review. *Archiwum Rybactwa Polskiego, 13*(1), 5.

Stępniewski, & Przywara (1992). The influence of soil oxygen availability on yield and nutrient uptake (N, P, K, Ca, Mg, Na) by winter rye (*Secale cereale*). *Plant and Soil, 143*(2), 267–274.

Stevens, Kader, & Albright (1979). Potential for increasing tomato flavor via increased sugar and acid content. *Journal of the American Society for Horticultural Science, 104*(1), 40–42.

Strayer (1994). Dynamics of microorganism populations in recirculating nutrient solutions. *Advances in Space Research, 14*(11), 357–366.

Sugita, Nakamura, & Shimada (2005). Microbial communities associated with filter materials in recirculating aquaculture systems of freshwater fish. *Aquaculture, 243*(1), 403–409.

Suhl, Baganz, Kloas, Dannehl, Jobs, Scheibe, & Schmidt (2018a). The potential of double recirculating aquaponic systems for intensive tomato production. *accepted by Acta Horticulturae*.

Suhl, Dannehl, Baganz, Schmidt, & Kloas (2018b). An innovative suction filter device reduces nitrogen loss in double recirculating aquaponic systems. *Aquacultural Engineering, 82*(63–72).

Suhl, Dannehl, Kloas, Baganz, Jobs, Scheibe, & Schmidt (2016). Advanced aquaponics: Evaluation of intensive tomato production in aquaponics vs. conventional hydroponics. *Agricultural Water Management, 178*, 335–344.

Suhl, Dannehl, Zechmeister, Baganz, Kloas, Lehmann, . . . Schmidt (2018c). Prospects and challenges of double recirculating aquaponic systems (DRAPS) for intensive plant production. *Acta Horticulturae, 1227*, 449–456.

Summerfelt (1998). An integrated approach to aquaculture waste management in flowing water systems. *In Proceedings of the Second International Conference on Recirculating Aquaculture, Roanoke, VA, USA, 16–19 July*, 87–97.

Theunissen, Ndakidemi, & Laubscher (2010). Potential of vermicompost produced from plant waste on the growth and nutrient status in vegetable production. *International Journal of Physical Sciences, 5*(13), 1964–1973.

Thorarinsdottir. (2015). *Aquaponics guidelines*. Reykjavik, Iceland: Haskolaprent.

Tilman, Balzer, Hill, & Befort (2011). Global food demand and the sustainable intensification of agriculture. *Proceedings of the National Academy of Sciences, 108*(50), 20260–20264.

Tilman, Cassman, Matson, Naylor, & Polasky (2002). Agricultural sustainability and intensive production practices. *Nature, 418*(6898), 671–677.

Timmons, Ebeling, Wheaton, Summerfelt, & Vinci. (2010). *Recirculating aquaculture systems* (C. A. Ventures Ed. Vol. 2). Ithaca NY, USA: Timmons, M.B. & Ebeling, J.M.

Torrellas, Antón, López, Baeza, Parra, Muñoz, & Montero (2012). LCA of a tomato crop in a multi-tunnel greenhouse in Almeria. *The International Journal of Life Cycle Assessment, 17*(7), 863–875.

Tovar, Moreno, Mánuel-Vez, & Garcia-Vargas (2000). Environmental impacts of intensive aquaculture in marine waters. *Water Research, 34*(1), 334–342.

Trejo-Téllez, & Gómez-Merino. (2012). Nutrient solutions for hydroponic systems. In *Hydroponics-A standard methodology for plant biological researches*: InTech.

Tyson, Simonne, Treadwell, Davis, & White (2008). Effect of water pH on yield and nutritional status of greenhouse cucumber grown in recirculating hydroponics. *Journal of Plant Nutrition, 31*(11), 2018–2030.

Tyson, Simonne, White, & Lamb. (2004). *Reconciling water quality parameters impacting nitrification in aquaponics: The pH levels.* Paper presented at the Proc. Fla. State Hort. Soc.

Tyson, Treadwell, & Simonne (2011). Opportunities and challenges to sustainability in aquaponic systems. *HortTechnology, 21*(1), 6–13.

UBA. (2015a). Prozessorientierte Basisdaten für Umweltmanagement-Instrumente (ProBas) (database). Retrieved 08.12.2017, from Umweltbundesamt http://www.probas.umweltbundesamt.de/php/index.php

UBA. (2015b). Prozessorientierte Basisdaten für Umweltmanagementsysteme (ProBas) (Library). ProBas Retrieved 11.11.2017, from Umweltbundesamt, Internationale Institut für Nachhaltigkeitsanalysen und -strategien http://www.probas.umweltbundesamt.de/php/index.php

UBA. (2017). *Umweltschutz in der Landwirtschaft*. Retrieved from Dessau-Roßlau, Germany:

https://www.umweltbundesamt.de/themen/umweltschutz-in-der-landwirtschaft

Ultee, Souvatzi, Maniadi, & König (2004). Identification of the culturable and nonculturable bacterial population in ground water of a municipal water supply in Germany. *Journal of Applied Microbiology, 96*(3), 560–568.

van de Nieuwegiessen, Boerlage, Verreth, & Schrama (2008). Assessing the effects of a chronic stressor, stocking density, on welfare indicators of juvenile African catfish, *Clarias gariepinus* Burchell. *Applied Animal Behaviour Science, 115*(3), 233–243.

van de Nieuwegiessen, Olwo, Khong, Verreth, & Schrama (2009). Effects of age and stocking density on the welfare of African catfish, *Clarias gariepinus* Burchell. *Aquaculture, 288*(1), 69–75.

van Os, Holterman, & Klomp. (1993). *Management of emission flows of pesticides from glasshouses.* Paper presented at the Symposium on engineering as a tool to reduce pesticide consumption and operator hazards in horticulture 372.

van Rijn (2013). Waste treatment in recirculating aquaculture systems. *Aquacultural Engineering, 53*, 49–56.

Verdegem (2013). Nutrient discharge from aquaculture operations in function of system design and production environment. *Reviews in Aquaculture, 5*(3), 158–171.

Verdegem, Bosma, & Verreth (2006). Reducing water use for animal production through aquaculture. *Water Resources Development, 22*(1), 101–113.

Vergote, & Vermeulen (2012). Recirculation aquaculture system (RAS) with tilapia in a hydroponic system with tomatoes. *Acta Horticulturae, 927*, 67–74.

Vilbergsson, Oddsson, & Unnthorsson (2016). Taxonomy of means and ends in aquaculture production—Part 2: The technical solutions of controlling solids, dissolved gasses and pH. *Water, 8*(9), 387.

Villarroel, Junge, Komives, König, Plaza, Bittsánszky, & Joly (2016). Survey of aquaponics in Europe. *Water, 8*(10), 468.

Verordnung (EU) Nr. 1258/2011 der kommission vom 2. Dezember 2011 zur Änderung der Verordnung (EG) Nr. 1881/2006 bezüglich der Höchstgehalte für Nitrate in Lebensmitteln, L 320/15 C.F.R. (2011).

Waechter-Kristensen, Caspersen, Adalsteinsson, Sundin, & Jensen (1999). Organic compounds and micro-organisms in closed, hydroponic culture: Occurrence and effects on plant growth and mineral nutrition. *Acta Horticulturae, 481*, 197–204.

Wang, Ouyang, Liu, Zhu, Zhao, Bao, & Hu (2014). Fruit and vegetable consumption and mortality from all causes, cardiovascular disease, and cancer: Systematic review and dose-response meta-analysis of prospective cohort studies. *BMJ : British Medical Journal, 349*.

WHO. (1990). *Diet, nutrition, and the prevention of chronic diseases*. Retrieved from Geneva, Schwitzerland:

WHO. (2013). *Global action plan for the prevention and control of noncommunicable diseases 2013–2020*. Retrieved from Geneva, Switzerland:

Wongkiew, Hu, Chandran, Lee, & Khanal (2017). Nitrogen transformations in aquaponic systems: A review. *Aquacultural Engineering, 76*, 9–19.

Wood, & Cowie (2004). A review of greenhouse gas emission factors for fertiliser production. *For IEA Bioenergy Task 38*, 1–20.

worldometers.info. World population forecast. Retrieved 12.09.2017, from Worldometers.info worldometers.info

Wortman (2015). Crop physiological response to nutrient solution electrical conductivity and pH in an ebb-and-flow hydroponic system. *Scientia Horticulturae, 194*, 34–42.

Yamin, Borisover, Cohen, & van Rijn (2017). Accumulation of humic-like and proteinaceous dissolved organic matter in zero-discharge aquaculture systems as revealed by fluorescence EEM spectroscopy. *Water research, 108*, 412–421.

Yep, & Zheng (2019). Aquaponic trends and challenges–A review. *Journal of Cleaner Production*.

Zechmeister. (2017). *Einfluss verschiedener Filtersysteme auf die Stickstoffdynamik und Berechnungen zum Düngereinsparpotential unter Einfluss verschiedener Pflanzen- und Fischarten in der Aquaponik.* (Bachelor thesis), Humboldt-Universität zu Berlin, Berlin.

Zekki, Gauthier, & Gosselin (1996). Growth, productivity, and mineral composition of hydroponically cultivated greenhouse tomatoes, with or without nutrient solution recycling. *Journal of the American Society for Horticultural Science, 121*(6), 1082–1088.

Zelená, Holasová, Zelený, Fiedlerová, Novotná, Landfeld, & Houška (2009). Effect of sulphur fertilisation on lycopene content and colour of tomato fruits. *Czech Journal of Food Science, 27*, 80–84.

Zeroni, Gale, & Ben-Asher (1983). Root aeration in a deep hydroponic system and its effect on growth and yield of tomato. *Scientia Horticulturae, 19*(3–4), 213–220.

Zhang, Dou, He, Ju, Powlson, Chadwick, . . . Wu (2013). New technologies reduce greenhouse gas emissions from nitrogenous fertilizer in China. *Proceedings of the National Academy of Sciences, 110*(21), 8375–8380.

Zieslin, & Snir (1989). Responses of rose plants cultivar 'Sonia' and Rosa indica major to changes in pH and aeration of the root environment in hydroponic culture. *Scientia Horticulturae, 37*(4), 339–349.